Menas Kafatos   Robert Nadeau

The Conscious Universe

Parts and Wholes in Physical Reality

With 30 Illustrations

Springer

Menas Kafatos
Department of Physics
George Mason University
Fairfax, VA 22030-4444
USA

Robert Nadeau
Department of English
George Mason University
Fairfax, VA 22030-4444
USA

Original artwork for the cover and figure illustrations created by Menas Kafatos.

Library of Congress Cataloging-in-Publication Data
Kafatos, Menas C.
    The conscious universe : parts and wholes in physical reality /
Menas Kafatos, Robert Nadeau.
        p.   cm.
    Includes bibliographical references.
    ISBN 0-387-98865-3 (alk. paper)
    1. Reality.   2. Physics—Philosophy.   3. Quantum theory.
I. Nadeau, Robert, 1944–   . II. Title.
QC6.4.R42K34   1999
530′.01—dc21   99-15364

Printed on acid-free paper.

Production managed by MaryAnn Cottone; manufacturing supervised by Jeffrey Taub.
Photocomposed copy prepared using author-supplied WordPerfect files.
Printed and bound by R.R. Donnelley and Sons, Harrisonburg, VA.
Printed in the United States of America.

9 8 7 6 5 4 3 2 1

ISBN 0-387-98865-3 Springer-Verlag New York Berlin Heidelberg   SPIN 10728757

*To Alexios Kafatos, Langdon Nadeau, Lefteris Kafatos,
Stefanos Kafatos, and Thalia Kafatou*

Contents

9. The Ceremony of Innocence: Physics, Metaphysics, and the Dialog between Science and Religion   143

Appendix. Horizons of Knowledge in Cosmological Models   162

Notes   172

Index   181

Introduction

Imagine that two people have been chosen to be observers in a scientific experiment involving two photons, or quanta of light. These photons originate from a single source and travel in opposite directions an equal distance halfway across the known universe to points where each will be measured or observed. Now suppose that before the photons are released, one observer is magically transported to a point of observation halfway across the known universe and the second observer is magically transported to another point an equal distance in the opposite direction. The task of the observers is to record or measure a certain property of each photon with detectors located at the two points so that the data gathered at each can later be compared.

Even though the photons are traveling from the source at the speed of light, each observer would have to wait billions of years for one of the photons to arrive at his observation point. Suppose, however, that the observers are willing to endure this wait because they hope to test the predictions of a mathematical theorem. This theorem not only allows for the prospect that there could be a correlation between the observed properties of the two photons but also indicates that this correlation could occur instantly, or in no time, in spite of the fact that the distance between the observers and their measuring instruments is billions of light years. Now imagine that after the observations are made, the observers are magically transported back to the source of the experiment and the observations recorded by each are compared. The result of our imaginary experiment is that the observed properties of the two photons did, in fact, correlate with one another over this fast distance instantly, or in no time, and the researchers conclude that the two photons remained in communication with one another in spite of this distance.

This imaginary experiment distorts some of the more refined aspects of the actual experiments in which photons released from a single source are measured or correlated over what physicists term *space-like* separated regions. But if we assume that the imaginary experiment was conducted many times, there is good reason to believe that the results would be the same as those in the actual experiments. Also like the imaginary experiment, the actual experiments were designed to test some predictions made in a mathematical theorem.

The theorem was published in 1964 by physicist John Bell, and the predictions made in this theorem have been tested in a series of increasingly refined experiments. Like Einstein before him, John Bell was discomforted by the threats that quantum physics posed to a fundamental assumption in classical physics: that there must be a one-to-one correspondence between every element of a physical theory and the physical reality described by that theory. This view of the relationship between physical theory and physical reality assumes that all events in the cosmos can be fully described by physical laws and that the future of any physical system can, in theory at least, be predicted with utter precision and certainty. Bell's hope was that the results of the experiments testing his theorem would obviate challenges posed by quantum physics to this understanding of the relationship between physical theory and physical reality.

The results of these experiments would also serve to resolve other large questions. Is quantum physics a self-consistent theory whose predictions would hold in this new class of experiments? Or would the results reveal that quantum theory is incomplete and its apparent challenges to the classical understanding of the correspondence between physical theory and physical reality were illusory? But the answer to this question in the experiments made possible by Bell's theorem would not merely serve as commentary on the character of the knowledge we call physics. It would also determine which of two fundamentally different assumptions about the character of physical reality is correct. Is physical reality, as classical physics assumes, local, or is physical reality, as quantum theory predicts, nonlocal? Although the question may seem esoteric and the terms innocuous, the issues at stake and the implications involved are, as we shall see, enormous.

Bell was personally convinced that the totality of all of our previous knowledge of physical reality, not to mention the laws of physics, would favor the assumption of locality. The assumption states that a measurement at one point in space cannot influence what occurs at another point in space if the distance between the points is large enough so that no signal can travel between them at light speed in the time allowed for measurement. In the jargon of physics, the two points exist in space-like separated regions, and a measurement in one region cannot influence what occurs in the other.

Quantum physics, however, allows for what Einstein disparagingly termed "spooky actions at a distance." When particles originate under certain conditions, quantum theory predicts that a measurement of one particle will correlate with the state of another particle even if the distance between the particles is millions of light years. And the theory also indicates that even though no signal can travel faster than light, the correlations will occur instantaneously, or in no time. If this prediction held in experiments testing Bell's theorem, we would be forced to conclude that physical reality is nonlocal.

After Bell published his theorem in 1964, a series of increasingly refined tests by many physicists of the predictions made in the theorem culminated in experiments by Alain Aspect and his team at the University of Paris-South. When the results of the Aspect experiments were published in 1982, the answers to Bell's questions were quite clear: Quantum physics is a self-consistent theory and the character of physical reality as disclosed by quantum physics is nonlocal.

In 1997, these same answers were provided by the results of twin-photon experiments carried out by Nicolus Gisin and his team at the University of Geneva.[1] The Gisin experiments were quite startling. While the distance between detectors in space-like separated regions in the Aspect experiments was 13 meters, the distance between detectors in the Gisin experiments was extended to 11 kilometers, or roughly 7 miles. Because a distance of 7 miles is quite vast within the domain of quantum physics, these results strongly indicate that similar correlations would exist even if experiments could be performed where the distance between the points was halfway across the known universe.

For reasons that will become clear later, what is most perplexing about nonlocality from a scientific point of view is that it cannot be viewed in principle as an observed phenomenon. The observed phenomena in the Aspect and Gisin experiments reveal correlations between properties of quanta, light, or photons, emanating from a single source based on measurements made in space-like separated regions. What cannot be measured or observed in this experimental situation, however, is the total reality that exists between the two points. Although the correlations allow us to infer the existence of this whole, they cannot in principle disclose or prove its existence.

When we consider that all quanta have interacted at some point in the history of the cosmos in the manner that quanta interact at the origins in these experiments and that there is no limit to the number of correlations that can exist between these quanta,[2] this leads to another dramatic conclusion: that nonlocality is a fundamental property of the entire universe. The daunting realization here is that the reality whose existence is inferred between the two points in the Aspect and Gisin experiments is the reality that underlies and informs all physical events in the universe. Yet all that we can say about this reality is that it manifests as an indivisible or undivided whole whose existence is inferred where there is an interaction with an observer or with instruments of observation.

If we also concede that an indivisible whole contains, by definition, no separate parts and that a phenomenon can be assumed to be real only when it is an observed phenomenon, we are led to more interesting conclusions. The indivisible whole whose existence is inferred in the results of the Aspect and Gisin experiments cannot in principle be the subject of scientific investigation. There is a simple reason why this is the case. Science can claim knowledge of physical reality only when the predictions of a physical theory are validated by experiment. Because the indivisible whole in the Aspect and Gisin experiments cannot be measured or observed, we confront an event horizon of knowledge, where science can say nothing about the actual character of this reality. We will discuss why this is the case in detail later.

If nonlocality is a property of the entire universe, then we must also conclude that an undivided wholeness exists on the most primary and basic level in all aspects of physical reality. What we are actually dealing with in science per se, however, are manifestations of this reality that are invoked or actualized in making acts of observation or measurement. Because the reality that exists between the space-like separated regions is a whole whose existence can only be inferred in experiments, as opposed to proven, the correlations between the particles, or the sum of these parts, does not constitute the indivisible whole. Physical theory allows us to understand why the correlations occur. But it cannot in principle disclose or describe the actual character of the indivisible whole.

Although the discovery that physical reality is nonlocal made the science section of *The New York Times*, it was not front-page news, and it received no mention in national news broadcasts. On these few occasions where nonlocality has been discussed in public forums, it is generally described as a piece of esoteric knowledge that has meaning and value only in the community of physicists. The obvious question is why a discovery that many regard as the most momentous in the history of science has received such scant attention and stirred so little debate. One possible explanation is that some level of scientific literacy is required to understand what nonlocality has revealed about the character of physical reality. Another is that the implications of this discovery have shocked and amazed scientists, and a consensus view of what those implications are has only recently begun to emerge.

The implication that has most troubled physicists is that classical epistemology, also known as Einsteinian epistemology, and an associated view of the character of scientific epistemology, known as the doctrine of positivism, can no longer be considered valid. Classical or Einsteinian epistemology assumes that there must be a one-to-one correspondence between every element in the mathematical theory and every aspect of the physical reality described that by that theory. And the doctrine of positivism assumes that the meaning of physical theories resides only in the mathematical description, as opposed to any nonmathematical constructs associated with this description. For reasons that will soon become obvious, the doctrine of positivism is premised on classical or Einsteinian epistemology, and the efficacy of both has been challenged by results of experiments testing Bell's theorem.

The results of these experiments have also revealed the existence of a profound new relationship between parts (quanta) and whole (universe) that carries large implications. Our proposed new understanding of the relationship between part and whole in physical reality is framed within the larger context of the history of mathematical physics, the origins and extensions of the classical view of the foundations of scientific knowledge, and the various ways that physicists have attempted to obviate previous challenges to the efficacy of classical epistemology. We will demonstrate why the discovery of nonlocality forces us to abandon this epistemology and propose an alternative understanding of the actual character of scientific epistemology originally articulated by the Danish physicist Niels Bohr. This discussion will serve as background for understanding a new relationship between parts and wholes in a quantum mechanical universe and a similar view of that relationship that has emerged in the so-called new biology.

What may prove most significant in this discussion in more narrowly scientific terms are the two chapters on physical cosmology, or the study of the origins and history of the entire universe. According to Niels Bohr, the logical framework of complementarity is not only required to understand the actual character of physical reality; it is also, he claimed, the most fundamental dynamic in our conscious constructions of reality in the mathematical language of physical theories.

Drawing extensively on Bohr's definition of this framework and applying it to areas of knowledge that did not exist during his time, we will attempt to show that his thesis is correct. We will demonstrate that complementarity has been a primary feature in every physical theory advanced in mathematical physics beginning with the special theory of relativity in 1905. And we will make the case that complementarity is an emergent property or dynamic in the life of the evolving universe at increasingly larger scales and times and that new part-whole complementarities emerged at greater levels of complexity in biological life. Based on this evidence, we will advance the hypothesis that future advances in physical theory in cosmology, or in the study of the origins and evolution of the entire universe, will also feature complementary constructs.

We will also make a philosophical argument that carries large implications in human terms that may initially seem very radical. Based on our new understanding of the relationship between parts and wholes in physics and biology, we will argue that human consciousness can be viewed as an emergent phenomenon in a seamlessly interconnected quantum universe. And we will make the case that nonlocality allows us to reasonably infer, without being able to prove, that the universe is a conscious system that evinces self-organizing and self-regulating properties that result in emergent order. We will, however, take care in this discussion to distinguish between what can be proven in scientific terms and what can be reasonably inferred in philosophical terms.

Physics and Metaphysics

Because we are concerned with the relationship between the new physics and metaphysics, let us be clear at the outset about our view of the actual character of that relationship. Most popular books that explore this relationship argue that the world-view of modern physics is more consistent with eastern metaphysics, particularly Taoism, Hinduism, and Buddhism. Although some writers who have drawn parallels between modern physics and eastern metaphysics are well known and respected physicists, like Fritjof Capra and David Bohm, most physicists tend not to be terribly impressed by these efforts. The obvious explanation for this reaction is that most physicists are convinced that physics has nothing to do with metaphysics, and, therefore, that any attempt to force a dialog between science and religion can only result in dangerous and groundless speculation.

Although metaphysical assumptions have played a role in the history of science and continue to play this role in what we will term the "hidden ontology of classical epistemology," metaphysics in our view should have, ideally at least, nothing to do with the actual practice of physics. Yet we will also make the case that the discovery that nonlocality is a new fact of nature allows us to infer in philosophical terms, although certainly not to prove in scientific terms, that the universe can be viewed as a conscious system. What makes these seemingly incompatible and contradictory positions self-consistent requires some explanation. Before we provide that explanation, it is necessary to articulate our views on the character of scientific truths, on the manner in which the truths of mathematical physics evolve, and on the relation of these truths to those in other aspects of human experience.

In our view, science is a rational enterprise committed to obtaining knowledge about the actual character of physical reality. We also believe that the only way to properly study the history and progress of science is to commit oneself to metaphysical and epistemological realism. Metaphysical realism assumes that physical reality has an objective existence outside or prior existence to human observation or any acts of measurement. And epistemological realism requires strict adherence to and regard for the rules and procedures for doing science as a precondition for drawing any conclusions worthy of the name scientific.

In classical physics, metaphysical and epistemological realism were regarded as self-evident truths, and no physical theory was presumed valid unless its predictions were subject to proof in repeatable scientific experiments under controlled conditions. In quantum physics, however, these self-evident truths became problematic due to the threats posed by wave-particle dualism and quantum indeterminacy to the classical epistemology. As we shall see in more detail later, the physical theory that describes the wave aspect of a quantum system is classical in the sense that it allows us to assume a one-to-one correspondence between every element of the physical theory and physical reality. If we do not measure or observe a quantum system, we can assume, theoretically at least, that we can know with certainty the state of this system. But if the quantum system is measured or observed, we cannot predict with complete certainty where the particle aspect of this system will appear. We can only calculate the range within which the particle aspect will appear, and we cannot know in principle where it will actually appear.

In an attempt to preserve the classical view of one-to-one correspondence between every element of the physical theory and physical reality, some physicists have assumed that the wave aspect of a quantum system is real in the absence of observation or measurement. Based on this assumption, several well-known physicists have posited theories with large cosmological implications in an attempt to obviate or subvert wave-particle dualism and quantum indeterminacy. As we hope to demonstrate, however, Bell's theorem and the experiments testing that theorem have revealed that these attempts to preserve the classical view of correspondence are not in principle subject to experimental proof, and must, therefore, be viewed as little more than philosophical speculation.

When we properly evaluate the observational conditions and results of experiments testing Bell's theorem, it becomes clear that wave-particle dualism and quantum indeterminacy are facts of nature that must be factored into our understanding of the nature of scientific epistemology. In doing so, we are obliged to recognize that any phenomena alleged to exist in the absence of observation or measurement in quantum physics cannot be viewed as real. As physicist John Archibald Wheeler puts it, "no phenomenon can be presumed to be a real phenomenon until it is an observed phenomenon."

If one can accept, along with most physical scientists, these definitions of metaphysical and epistemological realism, most of the conclusions drawn here should appear fairly self-evident in logical and philosophical terms. And it is not necessary to attribute any extra-scientific properties to our new understanding of the relationship between parts (quanta) and whole (cosmos) to embrace the new view of human consciousness that is consistent with this relationship. But since the conditions and results of experiments testing Bell's theorem also reveal that science can say nothing about the actual character of this whole, science can neither prove or disprove that our view of the relationship between part (human consciousness) and whole (cosmos) is correct. One is, therefore, free to dismiss our proposed view of this relationship for the same reason one is free to embrace it—the existence of this whole can only be inferred and the actual relationship of human consciousness to this whole cannot be known or defined in scientific terms.

Science as a Way of Knowing

As previously noted, we will argue that the discovery of nonlocality obliges us to abandon the classical view of one-to-one correspondence between physical theory and physical reality and the associated doctrine of positivism. But this in no way comprises the privileged character of scientific knowledge. Modern physical theories have allowed us to understand the origins and history of physical reality at all scales and times and to predict future events in physical reality with remarkable precision and certainty. Yet there are many well-educated humanists and social scientists, including some philosophers of science, who have adopted assumptions about the character of scientific truths that serve to either greatly diminish their authority or, in the extreme case, to render these truths virtually irrelevant to the pursuit of knowledge.

Those who promote these views typically appeal to the work of philosophers of science, principally that of Toulmin, Kuhn, Hanson, and Feyerabend. All of these philosophers assume that science is done within the context of a *Weltanschauung*, or comprehensive world-view, which is a product of culture and constructed primarily in ordinary, or linguistically based, language. One would be foolish to discount this view entirely. But it can, if taken to extremes, lead to some rather untenable and even absurd conclusions about the progress of science and its epistemological authority.

Although some physicists have taken the views of the *Weltanschauung* theorists seriously, most physicists have not. This entire approach to the philosophy of science also appears to have been largely displaced by historical realism. This approach is characterized, says Frederick Suppe, by "paying close attention to actual scientific practice, both historical and contemporary, all in the aim of developing a systematic philosophical understanding of the justification of knowledge claims."[3] If we were to identify ourselves with the views of any figures in the philosophy of science who practice historical realism, that figure would be, with some reservations, Dudley Shapere.

Shapere's focus is on the reasoning patterns in actual science and on the manner in which physics as a "privileged" form of coordinating experience with physical reality has obliged us to change our views of ourselves and the universe. We also agree with Shapere's view that the cumulative progress of science imposes constraints on what can be viewed as a legitimate scientific concept, problem, or hypothesis, and that these constraints become "tighter" as science progresses. This is particularly true when the results of theory present us with radically new and seemingly counterintuitive findings like those of the Aspect experiments. It is because there is incessant feedback within the content and conduct of science that we are led to counterintuitive results like the discovery of nonlocality as a fact of nature.

We also agree with Shapere's claim that the postulates of rationality, generalizability, and systematizability are rather consistently vindicated in the history of science.[4] Shapere does not dismiss the prospect that theory and observation can be conditioned by extrascientific, linguistically based factors. But he also argues, correctly in our view, that this does not finally compromise the objectivity of scientific knowledge. Although the psychological and sociological context of the scientist is an important aspect of the study of the history and evolution of scientific thought, the progress of science is not ultimately directed or governed by such considerations. Why this is the case should become quite clear in the course of this discussion.

There is, of course, no universally held view of the actual character of physical reality or of the epistemological implications of quantum physics. It would be both foolish and arrogant to claim that we have articulated this view or resolved the debate about quantum epistemology. At the same time, we are convinced that the view of physical reality advanced here is quite consistent with the totality of knowledge in mathematical physics and biology and that our proposed resolution of epistemological dilemmas is very much in accord with this knowledge.

1
Two Small Clouds: The Emergence of a New Physics

> Some physicists would prefer to come back to the idea of an objective real world whose smallest parts exist objectively in the same sense as stones or trees exist independently of whether we observe them. That, however, is impossible.
>
> *Werner Heisenberg*

During the summer of 1900, David Hilbert, widely recognized for his ability to see mathematics as a whole, delivered the keynote address at the Mathematics Congress in Paris. Speaking in a hall a few blocks away from the laboratory in which Madame and Pierre Curie were tending their vats of radioactive material, Hilbert set the agenda for the study of mathematics in the twentieth century. There were, he said, twenty-three unsolved problems in mathematics and all of them were amenable to solution in the near future. As Hilbert put it, "There is always a solution. There is no ignoramus."[1]

As it turned out, Hilbert's prediction that we would soon see the logically coherent and self-referential whole of mathematics by eliminating internal inconsistencies and problems was not accurate. The first major indication that this might be the case was the failed effort to resolve Hilbert's tenth problem. The solution to this problem required a mathematical proof that a certain kind of equation was solvable and that the solution could be found in a finite number of steps. This proof was not found, and we now know in principle that it will never be found. The failed attempt to resolve this and other related mathematical enigmas eventually culminated in the realization that the nineteenth-century view of mathematics as a self-referential whole that could prove it's logical self-consistency could not be sustained.

A few years prior to Hilbert's keynote address in 1900, Lord Kelvin, one of the best known and most respected physicists at that time, commented that "only two small clouds" remained on the horizon of knowledge in physics. In other words, there were, in Kelvin's view, only two sources of confusion in our otherwise complete understanding of material reality. The two clouds were the results of the Michelson-Morley experiment, which failed to detect the existence of a hypothetical substance called the ether, and the inability of electromagnetic theory to predict the distribution of radiant energy at different frequencies emitted by an idealized radiator called the "black body." These problems seemed so small that some established physicists were encouraging those contemplating graduate study in physics to select other fields of scientific study where there was better opportunity to make original contributions to knowledge. What Lord Kelvin could not have anticipated was that efforts to resolve these two anomalies would lead to relativity theory and quantum theory, or to what came to be called the "new physics."

The most intriguing aspect of Kelvin's metaphor for our purposes is that it is visual. It implies that we see physical reality through the mathematical description of nature in physical theory and the character of that which is seen is analogous to a physical horizon that is uniformly bright and clear. Obstacles to this seeing, the "two clouds," are likened to visual impediments that will disappear when better theory allows us to see through or beyond them to the luminous truths that will explain and eliminate them. The assumption that the mathematical description of nature can disclose such truths is, however, dependent on the assumption that mathematics is a logically consistent and self-referential system that can prove itself. There is, therefore, an intimate connection between Kelvin's belief that we can see all truths in physical reality and Hilbert's belief that we can see into the whole of mathematics and resolve or clarify all seeming inconsistencies. And the ground on which both beliefs rest is what we will term the "hidden ontology" of classical epistemology.

One reason that Kelvin's metaphor would have seemed quite natural and appropriate is that the objects of study in classical physics, like planets, containers with gases, wires, and magnets, were visualizable. His primary motive for metaphor can be better understood, however, in terms of some assumptions about the relationship between the observer and the observed system and the ability of physical theory to mediate this relationship. Observed systems in classical physics were understood as separate and distinct from the mind that investigates them, and physical theory was assumed to bridge the gap between these two domains of reality with ultimate completeness and certainty.

It is also interesting that light was the primary object of study in the new theories that would displace classical physics. Light in western literature, theology, and philosophy appears rather consistently as the symbol for transcendent, immaterial, and immutable forms separate from the realm of sensible objects and movements. Attempts to describe occasions when those forms and ideas appear known or revealed also consistently invoke light as that aspect of nature most closely associated with ultimate truths.

In the eighteenth century, when Alexander Pope penned the line, "God said, Let Newton be! and all was Light," he anticipated no ambiguity in the minds of his readers. There was now, assumed Pope, a new class of ultimate truths, physical law and theory, which had been revealed to man in the person of Newton. One irony is that the study of the phenomenon of light in the twentieth century leads to a vision of physical reality that is not visualizable or cannot be constructed in terms of our normative seeing in everyday experience. Another is that attempts to describe the actual behavior of light undermined the view of mathematical physics as a self-referential and logically consistent system that exists outside of and in complete correspondence with the dynamics of physical reality.

Light and Relativity Theory

Equally interesting, studies of light have often been foundational to the cumulative and context-driven progress of science that led to the questions asked in Bell's theorem. The best known of these experiments is probably that conducted in 1887 by Albert A. Michelson and Edward W. Morley. The intent was to refine existing theory, in this case Maxwell's electromagnetic theory, and both scientists were terribly disappointed when the effort failed. Light in Maxwell's theory is visualizable as a transverse wave consisting of magnetic and electric fields that vary in magnitude and direction in ways that are perpendicular to each other and to the direction of propagation of the wave (see Fig. 1). This wave theory of light had been established since the early 1800s and was well supported in experiments on light in which behavior like interference and diffraction had been observed. Interference arises when two waves, like those produced when two stones fall on the surface of a pond, combine to form larger waves when the crests of the two waves coincide. It can also be observed when waves cancel one another out when the crest of one wave corresponds with the trough of another. Diffraction is a wave property evident when waves bend around obstacles, like when ocean waves go around a wave-breaker in a harbor. Perhaps the best way to observe interference and refraction is to listen to sounds of musical notes, associated with sound waves, on a piano. Some combine and become louder while others cancel each other out.

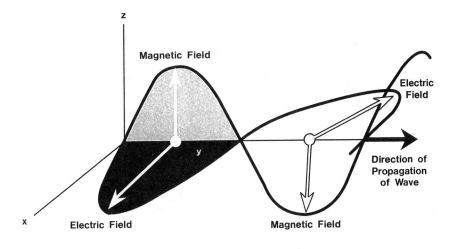

FIGURE 1. Light as an electromagnetic wave.

Because all known wave phenomena propagate through a material medium, it was natural to assume that light, which was viewed as electromagnetic waves, required a material medium through which its vibrant energy could propagate as well. The visualizable material medium whose existence was implied in the visualizable theory was, however, only a hypothesis, and Michelson and Morley were attempting to prove experimentally that the hypothetical medium, called the ether, was actually there.

According to classical theory, the ether would have to fill all of space, including the vacuum, and evince the stiffness of a material much stiffer than steel. Yet Michelson and Morley were convinced that something with these remarkable properties could be detected if an appropriate experiment was set up. What is suggested in their conviction that the experimental results would be positive is not naivete, but rather how complete the classical description appeared to physicists at the end of the nineteenth century.

In the Michelson-Morley experiment (see Fig. 2), a new device, called an interferometer, allowed accurate measurement of the speed of a beam of light. An original beam from a light source was split into two beams by a half-silvered mirror, and each beam was allowed to travel an equal distance along their respective paths. One beam was allowed to move in the direction that Earth moves, the other at 90 degrees with respect to the first.

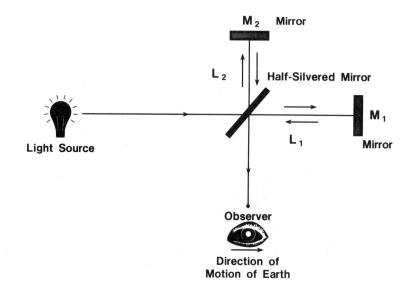

FIGURE 2. The Michelson-Morley experiment.

Based on the assumption that the hypothetical ether was absolutely at rest, the prediction was that the beam moving in the direction of Earth's movement would travel faster as it traveled through the ether due to the increase in velocity provided by the motion of Earth. Because that increase in velocity would not be a factor for the beam moving at 90 degrees with respect to the first, the expected result was that the interferometer would show a difference in the velocity of the two beams and confirm the actual existence of the ether. When no difference was found in the velocity of the beams, this result, which seemed as strange to Michelson and Morley as nonlocality seems strange in the experiments testing Bell's theorem, clearly indicated that the speed of light is constant. Although Einstein's relativity theory had not yet been invented to account for this result, that theory would eventually explain it.

At the age of sixteen, Einstein wondered what would happen, or what he would see, if he could somehow ride alongside a light wave.[2] Realizing in 1900 that the answer to this question required a thorough understanding of Maxwell's electrodynamic theory of light, he read the newest textbook on the subject by August Foppl and discovered that the electrodynamics of Maxwell depended on models that contradicted each other. There was one equation that described how a current is generated when a magnet is moving past a wire circuit and another that described how a current is generated when the wire circuit is moving past the magnet. Although some had speculated that the stationary ether could explain this difference, others argued that this did not make sense because the speed of light should also be affected by the presence of the ether. When Foppl addressed this issue, he referred to an experiment conducted by Hippolyte Fizeau in 1851 that showed that the speed of light was completely unaffected when running back and forth through a fast-moving column of water.

Einstein also read Poincare's *Science and Hypothesis,* a remarkably innovative book that was full of carefully reasoned and unsettling ideas. In a discussion of the relationship of geometry to space, Poincare wrote, "There is no absolute space" and "There is no absolute time."[3] He also said that "contradictions between competing theories" of the same phenomena "exist only as images we have formed to ourselves of reality" and that the "electrodynamics of moving bodies" suggest that the ether "does not actually exist."[4]

Sensing that Poincare's speculations may be correct, Einstein concluded that if the ether does not exist and the speed of light is constant, this requires some profound revisions of our understanding of the relationship between space and time. He was aware that the Newtonian construct of three-dimensional absolute space existing separately from absolute time implied that one could find a frame of reference absolutely at rest. And he knew that Newtonian mechanics also implied that it was possible to achieve velocities that corresponded to the speed of light and that the speed of light in this frame of reference would be reduced to zero.

Einstein's first postulate was that it is impossible to determine absolute motion, or motion that proceeds in a fixed direction at a constant speed. The only way, he reasoned, that we can assume such motion exists is to compare it with that of other objects. In the absence of such a comparison, said Einstein, one can make no assumptions about movement. He then concluded that the assumption that there is an absolute frame of reference in which the speed of light is reducible to zero must be false. Sensing that it was Newton's laws rather than Maxwell's equations that required adjustment, Einstein concluded that there is no absolute frame of reference or that the laws of physics hold equally well in all frames of reference. He then arrived at the second postulate of the absolute constancy of the speed of light for all moving observers. Based on these two postulates—the relativity of motion and the constancy of the speed of light—the entire logical structure of relativity theory followed.

Einstein mathematically deduced the laws that related space and time measurements made by one observer to the same measurements made by another observer moving uniformly relative to the first. Although Poincare had independently discovered the space-time transformation laws in 1905, he saw them as postulates without any apparent physical significance. Because Einstein perceived that the laws did have physical significance, he is recognized as the inventor of relativity. In this theory, the familiar law of simple addition of velocities does not hold for light or for speeds close to the speed of light. The reason these relativistic effects are not obvious in our everyday perception of reality, said Einstein, is that light speed is very large compared with ordinary speeds.

The primary impulse behind the special theory was a larger unification of physical theory that would serve to eliminate mathematical asymmetries apparent in existing classical theory. There was certainly nothing new here in the notion that frames of reference in conducting experiments are relative; Galileo had arrived at the same conclusion. What Einstein did, in essence, was extend the so-called Galilean relativity principle from mechanics, where it was known to work, to electromagnetic theory, or the rest of physics as it was then known. To achieve this greater symmetry, it was necessary to abandon the Newtonian idea of an absolute frame of reference and, along with it, the ether.

This led to the conclusion that the "electrodynamic fields are not states of the medium [the ether] and are not bound to any bearer, but they are independent realities which are not reducible to anything else."[5] He concluded that in a vacuum light traveled at a constant speed, c, equal to 300,000 km/sec, and thus all frames of reference become relative. There is, therefore, no frame of reference absolutely at rest, and the laws of physics could apply equally well to all frames of reference moving relative to each other.

Einstein also showed that the results of measuring instruments themselves must change from one frame of reference to another. Lorentz had earlier speculated that the reason the Michelson-Morley experiment did not detect differences in the speed of light was that the measuring apparatus was shrinking in the direction of the motion of Earth. He also developed his now famous transformation equations to translate the description of an event from one moving frame of reference to another. What Lorentz did not realize, however, was that "local time" was the only time that could be known in a moving frame of reference and that all other times were relative to it. Einstein realized this was the case and used the Lorentz equations to coordinate measurements in one frame of moving reference with respect to that in another frame. This means, for example, that clocks in the two frames of reference would not register the same time and two simultaneous events in a moving frame would appear to occur at different times in the unmoving frame.

For the observer in the stationary frame, lengths in the moving frame appear contracted along the direction of motion by a factor of $\sqrt{1 - v^2/c^2}$, *where v is the relative speed of the two frames. Masses, which provide a means to measure inertia, in the moving frame also appear larger to the stationary frame by the factor* $\dfrac{1}{\sqrt{1 - v^2/c^2}}$.

In the space-time description used to account for the differences in observation between different frames, time is another coordinate in addition to the three space coordinates forming the four-dimensional space-time continuum. In relativistic physics, transformations between different frames of reference express each coordinate of one frame as a combination of the coordinates of the other frame. For example, a space coordinate in one frame usually appears as a combination, or mixture, of space and time coordinates in another frame.

Entering the Realm of the Unvisualizable

It was the abandonment of the concept of an absolute frame of reference that began to move us out of the realm of the visualizable into the realm of the mathematically describable but unvisualizable. We can illustrate light speed with visualizable illustrations, like approaching a beam of light in a spacecraft at speeds fractionally close to that of light and imagining that the beam would still be leaving us at its own constant speed. But the illustration bears no relation to our direct experience with differences in velocity. It is when we try to image the four-dimensional reality of space-time as it is represented in mathematical theory that we have our first dramatic indication of the future direction of physics. It cannot be done no matter how many helpful diagrams and illustrations we employ.

As numerous experiments have shown, however, the counterintuitive results predicted by the theory of relativity occur in nature. For example, unstable particles, like muons, which travel close to the speed of light and decay into other particles with a well-known half-life, live much longer than their twin particles moving at lower speeds. Einstein was correct: The impression that events can be arranged in a single unique time sequence and measured with one universal physical yardstick is easily explained. The speed of light is so large compared with other speeds that we have the illusion that we see an event in the very instant in which it occurs.

To illustrate that simultaneity does not hold in all frames of reference, Einstein used a thought experiment featuring the fastest means of travel for human beings at his time—trains. What would happen, he wondered, if we were on a train that actually attained light speed? The answer is that lengths along the direction of motion would become so contracted as to disappear altogether and clocks would cease to run entirely. Three-dimensional objects would actually appear rotated so that a stationary observer could see the back of a rapidly approaching object. To the moving observer, all objects would appear to be converging on a single blinding point of light in the direction of motion.

Yet the train, as Einstein knew very well, could not in principle reach light speed. Only massless photons, or light, can reach light speed due to the equivalence of mass and energy. For a train, or spaceship, to reach this speed, mass would have to become infinite, and an infinite amount of energy would be required as well. While commonsense explanations of this situation may fail us, there is no ambiguity in the mathematical description.

Because light or photons have zero rest mass, they travel exactly at light speed. And in accordance with the Lorentz transformations, the factor $\sqrt{1 - v^2/c^2}$ *becomes zero as the relative speed v approaches light speed c.*

The special theory of relativity dealt only with constant, as opposed to accelerated, motion of the frames of reference, and the Lorentz transformations apply to frames moving with uniform motion with respect to each other. In 1915, Einstein extended relativity to account for the more general case of accelerated frames of reference in his general theory of relativity. The central idea in general relativity theory, which accounts for accelerated motion, is that it is impossible to distinguish between the effects of gravity and nonuniform motion. If we did not know, for example, that we were on an accelerating spaceship and dropped a cup of coffee, we could not determine whether the mess on the floor was due to the effects of gravity or the accelerated motion.

This inability to distinguish between a nonuniform motion, like an acceleration, and gravity is known as the principle of equivalence. This principle can also be interpreted as transforming the effects of gravity away as in free-falling frames of reference. In the physics of Einstein, the principle of equivalence explains the familiar phenomenon of astronauts floating in space as they circle Earth as a transformation of gravitational effects. This is quite different from the classical explanation that the effect is due to a balance between opposing forces.

In the general theory, Einstein posits the laws relating space and time measurements carried out by two observers moving uniformly, as in the example of one observer in an accelerating spaceship and another on Earth. Einstein concluded that force fields, like gravity, cause space-time to become warped (see Fig. 3) or curved and hence non-Euclidean in form. In the general theory the motion of material points, including light, is not along straight lines, as in Euclidean space, but along "geodesics" in curved space. The movement of light along curved spatial geodesics was confirmed in an experiment performed during a total eclipse of the sun by Arthur Eddington in 1919.

Here, as in the special theory, visualization may help to understand the situation but does not really describe it. This is nicely illustrated in the typical visual analogy used to illustrate what spatial geodesics mean. In this analogy we are asked to imagine a hypothetical flatland which, like a tremendous sheet of paper, extends infinitely in all directions. The inhabitants of this flatland, the flatlanders, are not aware of the third dimension. Because the world here is perfectly Euclidean, any measurement of the sum of the angles of triangles in flatland would equal 180 degrees, and any parallel lines, no matter how far extended, would never meet.

We are then asked to move our flatlanders to a new land on the surface of a large sphere. Initially, our relocated population would perceive their new world as identical to the old, or as Euclidean and flat. Next we suppose that the flatlanders make a technological breakthrough that allows them to send a kind of laser light along the surface of their new world for thousands of miles. The discovery is then made that if the two beams of light are sent in parallel directions, they come together after traveling a thousand miles.

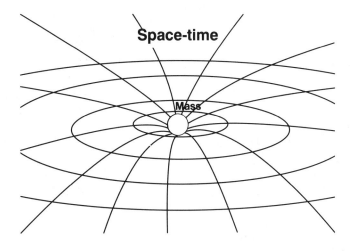

FIGURE 3. Warped space-time around a gravitating mass.

After experiencing utter confusion in the face of these results, the flatlanders eventually realize that their world is non-Euclidean, or curved like the surface of a sphere, and invent Riemannian geometry to describe the curved space. The analogy normally concludes with the suggestion that we are the flatlanders, with the difference being that our story takes place in three, rather than two, dimensions in space. Just as the shadow creatures could not visualize the curved two dimensional surface of their world, so we cannot visualize a three-dimensional curved space.

Thus a visual analogy used to illustrate the reality described by the general theory is useful only to the extent that it entices us into an acceptance of the proposition that the reality is unvisualizable. Yet here, as in the special theory, there is no ambiguity in the mathematical description of this reality. Although curved geodesics are not any more unphysical than straight lines, visualizing the three spatial dimensions as a "surface" in the higher four-dimensional space-time cannot be done. Visualization may help us better understand what is implied by the general theory, but it does not disclose what is really meant by the theory.

The Rise of Quantum Theory

The removal of Kelvin's second small cloud resulted in quantum theory and a description of physical reality that is even more unvisualizable than that disclosed by relativity theory. The first step was taken by German physicist Max Planck as he addressed the problem of the inability of current theory to explain black body radiation, and the object of study was, once again, light. A perfect black body absorbs all radiation that falls on it and emits radiant energy in the most efficient way as a function of its temperature. If you take, for example, a material object like a metal bar, put it in a dark, light-tight room, and heat it to a high temperature, it will produce a distribution of radiant energy with wavelength or colors that can be measured. If we make precise measurements of this radiation as the metal bar achieves higher temperatures and changes from dark red to white hot, a black body radiation curve that has a bell-shaped appearance can be obtained. The "cloud" here was that when the emission from all the vibrating charges was summed in accordance with electromagnetic theory: It predicted infinities as the frequency of light increased. This was clearly not in agreement with the observed bell-shaped behavior of black body intensity.

Kirchhoff, one of Planck's professors in graduate school, had earlier developed an unspecified function called the "black body" equation to describe what happens when a black body is heated and begins to emit radiation. He showed that the strength of the glow in any color was only a function of temperature regardless of the material used to create the black body. Kirchhoff then speculated that at any wavelength from blue to red the total amount of radiation energy emitted by a black body in a given time was a mathematical function of that wavelength and the absolute temperature. But he did not know what this function was, and this is the problem Planck began working on in 1900.

There were three black bodies in the German empire in 1900 and the newest of them could give off radiation that had never been measured before. The new black body used crystals of rock salt or fluorspar to reflect longer wavelengths and could measure radiation in the range where redness can no longer be seen and infrared begins. These experiments showed that although the intensity of the radiation characteristically peaked at shorter wavelengths, it came down from that peak at longer wavelengths. Planck wondered if an equation could relate the intensity of each wavelength to the temperature of the black body. But if the intensity at a given temperature peaked at a particular color and then came down again, the formula must feature a quantity that could describe this discontinuous process.

Planck visualized the mechanisms of the radiation of a black body as an enormous collection of tiny "oscillators" or resonators that could resonate with waves of certain frequencies, absorb these waves, and then emit them as radiation. Working with results of experiments done with the new black body at the Physikalisch-Technische Reichsanstalt in Berlin, Planck tackled this problem. And, like Michelson and Morley before him, he was not comfortable with the results. After failing to reconcile the results with existing theory, Planck concluded, in what he later described as "an act of sheer desperation," that the vibrating charges do not, as classical theory said they should, radiate light with all possible values of energy continuously. Based on the assumption that the material of the black body consisted of "vibrating oscillators," which would later be understood as subatomic events, he suggested that the energy exchange with the black body radiation is discrete or quantized.

Following this hunch, Planck viewed the energy radiated by a vibrating charge as an integral multiple of a certain unit of energy for that oscillator and found that the minimum unit of energy is proportional to the frequency of the oscillator. Working with this proportionality constant and calculating its value based on the careful data supplied by the experimental physicists, Planck solved the black body radiation problem. Although Planck could not have realized it at the time and would in ways live to regret it, his announcement of the explanation of black body radiation on December 14, 1900, was the birthday of quantum physics. Planck's new constant, known as the quantum of action, would later be applied to all microscopic phenomena. The fact that the constant is, like the speed of light, a universal constant would later serve to explain the strangeness of the new and unseen world of the quantum.

The next major breakthrough was made by the physicist who would eventually challenge the epistemological implications of quantum physics with the greatest precision and fervor. In the same year (1905) that the special theory appeared, Einstein published two other seminal papers that laid foundations for the revolution in progress. One was on the so-called Brownian movement, the other on the photoelectric effect.

In the paper on the photoelectric effect, Einstein challenged once again what had previously appeared in theory and experiment as obvious, and the object of study was, once again, light. The effect itself was a by-product of Heinrich Hertz's experiments, which at the time were widely viewed as having provided conclusive evidence that Maxwell's electromagnetic theory of light was valid. When Einstein explained the photoelectric effect, he showed precisely the opposite result—the inadequacy of classical notions to account for this phenomenon.

The photoelectric effect is witnessed when light with a frequency above a certain value falls on a photosensitive metal plate and ejects electrons (see Fig. 4). A photosensitive plate is one of two metal plates connected to ends of a battery and placed inside a vacuum tube. If the plate is connected to the negative end of the battery, light falling on the plate can cause electrons to be ejected from the negative end. These electrons then travel through the vacuum tube to the positive end and produce a flowing current.

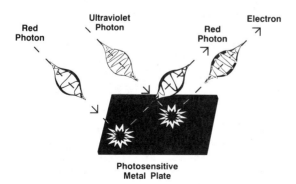

FIGURE 4. The photoelectric effect: A photon of low energy (red) cannot eject an electron but a photon of high-energy (ultraviolet) can.

In classical physics the amplitude, or height, of any wave, including electromagnetic waves, describes the energy contained in the wave. The problem Einstein sought to resolve can be thought about using water waves as an analogy. Large water waves, like ocean waves, have large height or amplitude, carry large amounts of energy, and are capable of moving many pebbles on a beach. Because the brightness of a light source is proportional to the amplitude of the electromagnetic field squared, it was assumed that a bright source of light should eject lots of electrons and that a weak source of light should eject few electrons. In other words the more powerful wave, the bright light, should move more pebbles, electrons, on this imaginary beach. The problem was that a very weak source of ultraviolet light was capable of ejecting electrons while a very bright source of lower-frequency light, like red light, could not. It was as if the short, choppy waves from the ultraviolet source could move pebbles, or electrons, on this imaginary beach, while the large waves from the red light source could not move any at all.

Einstein's explanation for these strange results was as simple as it was bold. In thinking about Planck's work on light quanta, he wondered if the exchange of energy also occurred between particles with mass, like electrons. He then concluded that the energy of light is not distributed evenly over the wave, as classical physics supposed, but is concentrated in small, discrete bundles. Rather than view light as waves, Einstein conceived of light as bundles, or "quanta," of energy in the manner of Planck. The reason that ultraviolet light ejects electrons and red light does not, said Einstein, is that the energy of these quanta is proportional to the frequency of light, or to its wavelength.

In this quantum picture, it is the energy of the individual quanta, rather than the brightness of the light source, that matters. While Planck had quantized only the interaction of matter with energy, Einstein quantized energy itself. Viewing the situation in these terms, individual red photons do not have sufficient energy to knock an electron out of the metal while individual ultraviolet photons have sufficient energy. When Einstein computed the constant of proportionality between energy and the frequency of the light, or photons, he found that it was equal to Planck's constant.

A New View of Atoms

The discovery of the element polonium, by Pierre and Madame Curie in 1898, had previously suggested that atoms were composite structures that transformed themselves into other structures as a result of radioactivity. It was, however, Einstein's paper on Brownian motion that finally enticed physicists to conceive of atoms as something more than a philosophical construct in the manner of the ancient Greeks. The motion is called "Brownian" after the British botanist Thomas Brown, who discovered in 1827 that when a pollen grain floating on a drop of water is examined under a microscope, it appears to move randomly. Einstein showed that this motion obeys a statistical law and the pattern of motion can be explained if we assume that objects, like pollen grains, are moving about as they collide at the microscopic level with tiny molecules of the water. Although Einstein did suggest that molecules and the atoms that constitute them were real in that their behavior had concrete effects on the macro level, nothing of substance was known at the time about the internal structure of atoms.

The suggestion that the world of the atom had a structure enticed Ernest Rutherford in Manchester to conduct a series of experiments in which positively charged "alpha" particles, later understood to be the nuclei of helium atoms, were emitted from radioactive substances and fired at a very thin sheet of gold foil. If there was nothing to impede the motion of the particles, they should travel in a straight line and collide with a screen of zinc sulfide where a tiny point of light, or scintillation, would record the impact.

In this experiment most of these particles were observed to be slightly deflected from their straight-line path. Other alpha particles, however, were deflected backward toward the direction from which they came. Based on an estimate of the number of alpha particles emitted by a gram of radium in one second, Rutherford was able to arrive at a more refined picture of the internal structure of the atom.

The existing model, invented by the discoverer of the electron, J.J. Thomson, presumed that the positive charge was distributed over the entire space of the atom. The observed behavior of the alpha particles suggested, however, that the particles deflected backward were encountering a highly concentrated positive charge, while most particles traveled through the space of the atoms as if this space were empty. Rutherford explained the results in terms of a picture of the atom as being composed primarily of vast regions of space in which the negatively charged particles, electrons, move around a positively charged nucleus that contains the greatest part of the mass of the atom.

Forced to appeal to macro-level analogies to visualize this unvisualizable structure, Rutherford termed the model "planetary." It was soon discovered, however, that there is practically no similarity between the structure or behavior of macro and micro worlds. The relative distances between electrons and nucleus, as compared to the size of the nucleus, are much greater than the relative distances between planets and the sun, as compared to the size of the sun. If one can imagine Earth undergoing a quantum transition and instantaneously appearing in the orbit of Mars, this illustrates how inappropriate macro-level analogies would soon become.

The next step on the road to quantum theory was made by a Danish physicist from whom we will hear a great deal more later in this discussion—Niels Bohr. Developed partly as a result of the work done with Rutherford in Manchester, Bohr provided, in a series of papers published in 1913, a new model for the structure of atoms. Although obliged to use macro-level analogies, Bohr was the first to suggest that the orbits of electrons were quantized. His model was semiclassical in that it incorporated ideas from classical celestial mechanics about orbiting masses. The problem he was seeking to resolve had to do with the spectral lines of hydrogen, which showed electrons occupying specific orbits at specific distances from the nucleus with no in-between orbits.

Spectral lines are produced when light from a bright source containing a gas, like hydrogen, is dispersed through a prism, and the pattern of the spectral lines is unique for each element. The study of the spectral lines of hydrogen suggested that the electrons somehow "jump" between the specific orbits and appear to absorb or emit energy in the form of light or photons in the process. What, wondered Bohr, was the connection?

Bohr discovered that if you use Planck's constant in combination with the known mass and charge of the electron, the approximate size of the hydrogen atom could be derived. Assuming that a jumping electron absorbs or emits energy in units of Planck's constant, in accordance with the formula Einstein used to explain the photoelectric effect, Bohr was able to find correlations with the specific spectral lines for hydrogen. More important, the model also served to explain why the electron does not, as electromagnetic theory says it should, radiate its energy quickly away and collapse into the nucleus.

Bohr reasoned that this does not occur because the orbits are quantized— electrons absorb and emit energy corresponding to the specific orbits. Their lowest energy state, or lowest orbit, is the "ground state" (see Fig. 5). What is notable here is that Bohr, although obliged to use macro-level analogies and classical theory, quickly and easily posits a view of the dynamics of the "energy shells" of the electron that has no macro-level analogy and is inexplicable within the framework of classical theory.

The central problem with Bohr's model from the perspective of classical theory was pointed out by Rutherford shortly before the first paper describing the model was published. "There appears to me," Rutherford wrote in a letter to Bohr, "one grave problem in your hypotheses which I have no doubt you fully realize, namely, how does an electron decide what frequency it is going to vibrate at when it passes from one stationary state to another? It seems to me that you would have to assume that the electron knows beforehand where it is going to stop."[6] Viewing the electron as atomic in the Greek sense, or as a point-like object that moves, there is cause to wonder, in the absence of a mechanistic explanation, how this object instantaneously "jumps" from one shell or orbit to another. It was essentially efforts to answer this question that led to the development of quantum theory.

The effect of Bohr's model was to raise more questions than it answered. Although the model suggested that we can explain the periodic table of the elements by assuming a maximum number of electrons are found in each shell, Bohr was not able to provide any mathematically acceptable explanation for the hypothesis. That explanation was provided in 1925 by Wolfgang Pauli, known throughout his career for his extraordinary talents as a mathematician.

Bohr had used four quantities in his model: Planck's constant, mass, and charge and a quantum number indicating the orbital level. Pauli added an additional quantum number, described as "spin," which was initially represented with the macro-level analogy of a spinning ball on a pool table. Rather predictably, the analogy does not work. Whereas a classical spin can point in any direction, a quantum mechanical spin points either up or down along the axis of measurement (see Fig. 6). In total contrast to the classical notion of a spinning ball, we cannot even speak of the spin of the particle if no axis is measured.

When Pauli added the spin quantum number, he found a correspondence between the number of electrons in each full shell of atoms and the new set of quantum numbers describing the shell. This became the basis for what we now call the Pauli exclusion principle. The principle is simple and yet quite startling: Two electrons cannot have all their quantum numbers the same, and no two actual electrons are identical in the sense of having the same quantum number. The exclusion principle explains mathematically why there is a maximum number of electrons in the shell of any given atom. If the shell is full, adding another electron would be impossible because this would result in two electrons in the shell having the same quantum numbers.

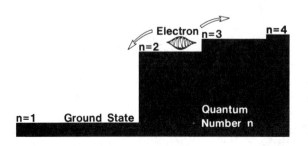

FIGURE 5. The energy levels in the Bohr atom can be visualized as a set of steps of different heights. The electron, visualized here as a wave packet, is always constrained to be found on one of the steps.

FIGURE 6. Quantization of spin:
Along a given direction in space, the
measured spin of an electron can
have only two values.

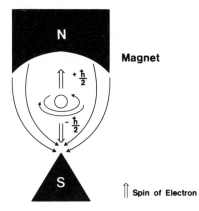

This may sound a bit esoteric, but the fact that nature obeys the exclusion principle is quite fortunate from our point of view. If electrons did not obey the principle, all elements would exist at the ground state and there would be no chemical affinity between them. Structures like crystals and DNA would not exist, and the only structures that would exist would be spheres held together by gravity. The principle allows for chemical bonds that, in turn, result in the hierarchy of structures from atoms, molecules, cells, plants, and animals.

Waves as Particles and Particles as Waves

The next development in the road toward quantum theory was based on experiments conducted by Arthur Compton using X-rays, and the results were published in 1923. Compton found that in a collision of an X-ray photon with an electron the total momentum of the system is conserved, and the wavelength of light changes appropriately. The results suggested that the photons were behaving like particles. If light had particle properties, when it was previously conceived as a wave, then perhaps the electron, previously conceived as a particle, also had wave properties. A French doctoral student in physics, Louis de Broglie, suggested in his thesis that the same formula Einstein applied to photons and Compton applied to the collisions of photons with electrons might also apply to all known particles. This came to be known as the de Broglie wavelength.

The existence of the so-called *"matter waves"* was demonstrated in experiments involving the scattering of electrons off crystals where electrons showed interference patterns indicative of wave properties. The consensus would eventually become that particles possess wave-like properties in the same way that light possesses particle-like properties. Thus de Broglie's hunch led to a large and unexpected uni-fication. It provided an explanation for the previously unexplained assertion in Bohr's model that an electron is confined to specific or-bits. An electron, concluded de Broglie, is confined to orbits in terms of integer numbers of waves.

De Broglie's thesis was brought to the attention of Einstein, who then brought it to the attention of Erwin Schrödinger, a professor in Zürich. Drawing on his classical understanding of wave phenomena, Schrödinger developed wave mechanics (1925). The nineteenth-century physicist William Hamilton had created a series of equations describing the geometrical particle-like and wave-like properties of light. Drawing on Hamilton's equations, Schrödinger assumed the "reality" of "matter waves" and described them with a wave function.

We now know that wave mechanics describes only one aspect of the total reality of a quantum system. The first insight that began to open the door to this improved understanding came from Max Born in 1926, and it was not well received by the majority of physicists at the time. Born realized that because the wave function itself cannot be observed, it is not a "real" entity in the classical sense. While the square of the wave function may give us the probability of finding a particle within a region, it does not, concluded Born, allow us to pre-cisely predict where that particle will be found.

What greatly disturbed physicists was that Born's definition of the term "probability" did not refer to a convenient way of assessing the overall behavior of a system that could, in theory, be described in classical terms. He was referring to an "inherent" aspect of measure-ment of all quantum mechanical events, which does not allow us to predict precisely where a particle will be observed no matter what improvements are made in experiments. While the quantum recipe that describes this situation is simple mathematically, the reality it describes is totally unvisualizable. The wave function is unobserv-able, and yet the square of the wave function gives us the probability of finding the particle within a particular region of space with certain properties.

Physicists compute the absolute value or amplitude of a wave by squaring its wave function, $|\Psi^2|$. The wave function defines the possi-bilities, and the experimental results are only predictable in terms of probabilities, i.e., probability $= |possibility|^2$.

The wave function provides a complete description of the quantum particle or system, and wave mechanics, in this sense, is a "complete" theory. In practice or actual experiments, however, the theory de-scribes only probabilities of events happening rather than specific events. The specific event cannot be predicted; what we can predict is only the probability that it may happen.

Einstein characterized the strangeness of this situation from a classical point of view by referring to the wave function as a "ghost field." Rather than representing a real matter wave, the wave function describes, suggested Einstein, only a wavy, probabilistic reality. Although this situation may seem simple enough mathematically, the real existence of wave and particle aspects of reality presented a direct challenge to the efficacy of classical or Einsteinian epistemology and the doctrine of positivism.

In 1925, the same year that Schrödinger was developing wave mechanics, Werner Heisenberg, Max Born, and Pascual Jordan were constructing an alternative set of rules for calculating the frequencies and intensities of spectral lines. Operating on the assumption that science can only deal in quantities that are measurable in experiments, their focus was on the particle aspect. The result was an alternative theoretical framework for quantum theory known as matrix mechanics.

Matrices involve calculations with a curious property: When two matrices are multiplied, the answer that we get depends on the order of their multiplication. In other words, for matrices, 2 × 3 would not be equal to 3 × 2, or in the language of algebra, a× b may not be equal to b × a. The word "matrix" is used here because in the Heisenberg-Borg-Jordan formulation of quantum theory, the alternative set of rules applied to organizing data into mathematical tables, or matrices. These tables were used to calculate probabilities associated with initial conditions that could be applied in the analysis of observables. As Heisenberg would reflect later, we have now arrived at the point where we must "abandon all attempts to construct perceptual models of atomic process."[7]

It is also significant that the point at which we fully enter via mathematical theory the realm of the unvisualizable is the point at which macro-level or classical logic breaks down. As Max Jammer, the recognized authority on the history of quantum mechanics, puts it, "It is hard to find in the history of physics two theories [wave and matrix mechanics] designed to cover the same range of experience which differ more radically than these two."[8] Heisenberg characterized his view of the situation with the analogy that it is as if a box were "full and empty at the same time."[9]

The confusion arises in part because of the classical assumption that all properties of a system, including those of microscopic atoms and molecules, are real in the sense that they are exactly definable and determinable. But, as Bohr was among the first to realize, in the quantum world positions and momenta (where "momentum" is defined as the product of mass times velocity) cannot be said to have definite values even in principle. Rather, we deal in probabilities, which in the Born formalism are expressed by the square of the amplitude of the wave function.

This role of observation of quantum systems not only challenged the classical view of the relationship between physical theory and physical reality, it also challenged the classical assumptions that the observer was separate and distinct from the observed system and that acts of observation did not alter the system. In quantum physics, a definite value of a physical quantity can be known only through acts of observation, which includes the observer and his measuring instruments, and we cannot assume that the quantity would be the same in the absence of observation. Put differently, we cannot assume that a physical system exists in a well-defined state prior to measurement or that this state will be the same when a measurement is made. Even if our predictions are based on complete knowledge of initial conditions, the future state of this system cannot be entirely predicted.

Werner Heisenberg responded to the new situation with his famous indeterminacy principle. The principle states that the product of the uncertainty in measuring the momentum, p, of a quantum particle times the uncertainty in measuring its position, x, is always greater than or equal to Planck's constant.

A comment by Robert Oppenheimer illustrates how bizarre this situation seemed in terms of normative or everyday logic: "If we ask, for instance, whether the position of the electron remains the same, we must say 'no'; if we ask whether the electron is at rest, we must say 'no'; if we ask whether it is in motion, we must say 'no.' "[10] We would soon realize that normative or everyday logic, which is premised on Aristotle's law of excluded middle, is based on our dealings with macro-level phenomena and does not hold in the quantum domain. It is this realization that led Bohr to develop his new logical framework of complementarity.

At this point in the history of modern physics, physicists divided into two camps. Planck, Schrödinger and de Broglie joined ranks with Einstein in resisting the implications of quantum theory. Figures like Dirac, Pauli, Jordan, Born, and Heisenberg became, in contrast, advocates of the Copenhagen interpretation of quantum mechanics. Meanwhile, quantum mechanics continued to be applied with remarkable success in its new form: quantum field theory. Here we witness the same correlation between increasingly elaborate mathematical descriptions of reality, a vision of the cosmos that is not visualizable, and the emergence of additional constructs that can only be understood within Bohr's new logical framework of complementarity.

The New Logical Framework of Complementarity

This new logical framework, which will assume increasingly more importance in this discussion, is a central feature of Bohr's Copenhagen Interpretation. It is this interpretation that is considered the "orthodox," or standard, interpretation by experts on the quantum measurement problem and quantum epistemology. Some physicists have chosen to include in their understanding of the Copenhagen Interpretation Born's commentary on the probability postulate and Heisenberg's idea of quantum potential. This results, however, in a radical distortion of what Bohr's orthodox interpretation actually means. As we shall see, Bohr confronts and resolves the epistemological implications of the quantum observation problem in utterly realistic terms. But because Bohr's interpretation forces us to question some cherished assumptions in classical epistemology, the logical framework of complementarity is generally not well understood by physical scientists.

As the physicist and philosopher of science Clifford Hooker notes, "Bohr's unique views are almost universally either overlooked completely or distorted beyond all recognition—this by philosophers of science and scientists alike."[11] Part of the explanation for this situation is that physicists begin their studies with classical mechanics, where classical epistemology is implicit, and receive little exposure to epistemological problems in their study of quantum physics. And because physicists are not obliged to think about epistemological problems in practical everyday applications of quantum theory, many continue to believe in classical epistemology despite the fact that a proper understanding of the conditions and results of their experiments would undermine their faith in this epistemology.

This explains why most physicists are troubled by Bohr's conclusion in the orthodox Copenhagen Interpretation that the truths of science are not, as the architects of classical physics believed, "revealed" truths. They are subjectively based constructs that are useful to the extent that they help us coordinate greater ranges of experience with physical reality. But this does not mean, as some have supposed, that Bohr took the position that the truths of science in physical theory are, in any sense, arbitrary. It is quite clear, as he often pointed out, that they coordinate our experience with physical reality beautifully and with great precision. Most physical scientists have tended to relegate Bohr's views to a file drawer called "Philosophy" in the hope that they will be obviated by further progress in physical theory and experiments. But this has not, in fact, occurred, and Bell's theorem and the experiments testing that theory clearly indicate that we must open that drawer and review its contents.

In the next chapters on the quantum mechanical view of nature and on Bell's theorem and the experiments testing it, we will continue our journey into the strange new world of quantum physics. The entrance fee for the uninitiated is a willingness to free oneself of the constraints of everyday visualizable reality and to freely exercise the imagination. Although this brave new world may seem, initially at least, bizarre, it represents, from a scientific point of view, the way things are.

2
The Strange New World of the Quantum: Wave-Particle Dualism

> The paradox is only a conflict between reality and your feeling of what reality ought to be.
>
> *Richard Feynman*

When Heisenberg said of wave-particle dualism that it was as if a box were "full and empty at the same time," he chose this Zen-like analogy not only because this strange dualism defies the normative Western sense of logical relationships; he was also suggesting that the allegedly "full" description of this reality in physical theory, wave mechanics, was "empty" in the sense that only the wave or particle aspect can be revealed in single acts of observation. Both aspects are never present in a single measurement, and the aspect that is apparent is a function of experimental choice. The new logic that describes this situation, developed by Niels Bohr, is known as the "logical framework of complementarity."

Most physicists were willing to accept this logic as a heuristic, or something that could be assumed without contradicting what was known. In Bohr's view, however, the logic of complementarity is not a heuristic per se; it is an emergent property of physical reality that must be factored into our attempts to coordinate experience with this reality in physical theory. This view has not been widely embraced in the community of physicists for a simple reason. If complementarity is an inherent aspect of physical reality on a very fundamental level, this would undermine the efficacy of classical or Einsteinian epistemology and the doctrine of positivism.

The essential paradox of wave-particle dualism is easily demonstrated. View the particle as a point-like something, like the period at the end of this sentence, and the wave as continuous and spread out (see Fig. 7). The obvious logical problem is how a particular something localized in space and time, the particle, can also be the spread-out and continuous something, the wave. Quantum physics not only says unequivocally that quanta exhibit both properties, it also provides mathematical formalism governing what we can possibly observe when we coordinate our experience with this reality in actual experiments.

In quantum physics, observational conditions and results are such that we cannot assume that there is a categorical distinction between the observer and the observing apparatus or between the mind of the physicist and the results of physical experiments. The measuring apparatus and the existence of an observer are essential aspects of the act of observation. What has consistently troubled physicists about this situation is that it implies that we can no longer see the pre-existent truths of physical reality through the lenses of physical theory in the classical sense.

The wave aspect of quanta, which may be crudely and inappropriately visualized as water waves, is responsible for the formation of interference patterns (see Fig. 8). Interference becomes apparent when two waves combine and produce a larger wave or when they cancel each other where the height of one is equal to the trough of the other.

In quantum physics, the wave function allows us to theoretically predict the future of a quantum system with complete certainty so long as the system is not observed or measured. But when an observation or measurement occurs, the wave function does not allow us to predict precisely where the particle will appear at a specific location in space. It only allows us to predict the probability of finding the particle within a range of probabilities associated with all possible states of the wave function.

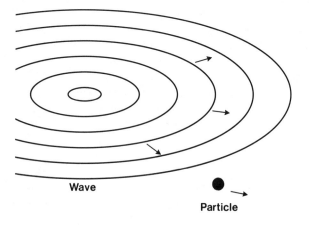

FIGURE 7. Wave and particle.

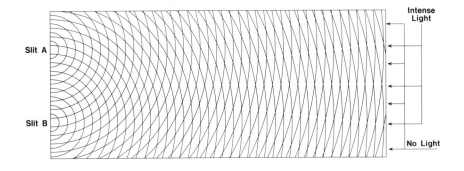

FIGURE 8. Drawing showing interference effects of two waves originating in slits A and B.

Because some of the probabilities associated with the wave function that seem real in the absence of observation are realized when the particle is observed and others are not, the wave function, in the jargon of physics, is said to "collapse" to one set of probabilities. The quantum strangeness here is that all the probabilities that seem to actually exist in the absence of observation are not realized when an observation occurs. One of the fundamental problems dealt with in quantum mechanics is to indicate where within the wave aspect of this reality we can expect to observe its particle aspect.

The "total reality" of a quantum system is wave and particle, and Bohr was among the first to realize that a proper understanding of the relationship between these two aspects of a single reality requires the use of a new logical framework. What makes the logical framework of complementarity new, or where it extends itself beyond our usual understanding of logical oppositions, is the following stipulation: In addition to representing profound oppositions that preclude one another in a given situation, both constructs are necessary to achieve a complete understanding of the entire situation. In other words, it is both logically disparate constructs that describe the total reality, even though only one can be applied in any given instance. Put in terms of Heisenberg's analogy, while our dealings with wave-particle dualism are empty in the sense that we cannot simultaneously disclose both of its complementary aspects, the total reality is always full in the sense that both aspects constitute the whole.

Wave mechanics describes the continuous movement in time of a multidimensional spread-out wave and is completely deterministic. This aspect of quantum mechanics is complete in the classical sense in that it describes everything that can possibly be known about the quantum system in the absence of observation. If we calculate the

possibilities given by the wave function and are not required to demonstrate that all these possibilities can be disclosed in a single experimental situation, wave mechanics appears to be the conceptual
lens that allows us to see into the essences of this reality.

The initial impulse of Schrödinger, de Broglie, and others was to
view the wave function as an actually existing entity, like water
waves. This classical view of the wave as having a separate and independent existence from the particle obviously allows one to assume
that there is complete correspondence between the physical theory
and physical reality in the classical sense. The problem became, however, that although the wave function theoretically describes everything that can possibly happen in a quantum system, the actual observation of the system must deal in only the probability of finding a
"something," or a quantum, at specific locations in space and in a specific energy state.

Much of the confusion about these disparate views of a single reality results from the fact that one description of this reality is classical. If a quantum system is left alone, meaning we do not attempt to
observe it, the properties of the system can be assumed to change
causally in accordance with the deterministic wave equation, like a
system in classical physics.[1] And yet the other aspect of this reality,
which is invoked when a measurement of the system is made, suggests that change in the system is discontinuous in accordance with
the laws of probability theory. Heisenberg's matrix mechanics and
Feynman's integral path approach represent two attempts to mathematically describe this aspect of the total reality.

As physicist Eugene P. Wigner has emphasized, attempts to describe wave and particle aspects of a quantum system reveal the most
fundamental dualism encountered in quantum theory.[2] On the one
hand, we have a classical system featuring unrestricted causality and
complete correspondence between every element in the physical theory and physical reality. On the other, we have a completely nonclassical system that features discontinuous processes, the absence of unrestricted causality, and the lack of a complete correspondence
between physical theory and physical reality.

The confusion has been amplified by the choice of the phrase "collapse of the wave function" to describe a situation where observation
or measurement of a quantum system occurs. The choice is unfortunate because it implies that the wave function, as the term matter-
wave initially suggested, is a real or actual thing that exists before
the act of observation or in the absence of observation.[3] Viewing the
wave function in this way requires that we assume that some aspects
of this system, which were real or actual in the absence of observation, somehow collapse or "disappear" when observation occurs. The
quantum formalism in Bohr's orthodox Copenhagen Interpretation

says nothing of the kind. What this formalism indicates is that before measurement we only have a range of possibilities given by the wave function.

These possibilities are mathematically derivable probabilities given by the square of its amplitude $|\Psi|^2$ *(see Fig. 9). When an actual measurement is made, or when something definite is recorded by our instruments, the various possibilities become one actuality.*

The wave equation of Schrödinger, which describes the evolution in space and time of the wave function in a totally deterministic fashion, cannot tell us what will actually occur when the system is observed. All the wave function can provide is a description of the range within which the particle aspect may be observed.

What has troubled physicists is that one aspect of this reality, as described in physical theory, suggests that we have a complete theory that mirrors the behavior of physical reality. However, our efforts to coordinate experience with the total reality requires the use of other, logically disparate, mathematical descriptions. From the perspective of classical epistemology, the problem is that an allegedly complete physical theory, quantum mechanics, does not and cannot allow us to describe when and how the collapse of the wave function occurs.

In Bohr's orthodox interpretation, the wave function is viewed merely as a mathematical device or idealization of a reality that cannot be directly measured or observed. The function expresses the relationship between the quantum system, which is inaccessible to the observer, and the measuring device, which conforms to classical physics. What seems confusing here, particularly given the fact that we live in a quantum universe, is the requirement that we view quantum reality with one set of assumptions, those of quantum physics, and the results of experiments where this reality is measured or observed with another set of assumptions, those of classical physics. This implies a categorical distinction between the micro and macro worlds, and yet does not specify at what point a measuring device ceases to be classical and becomes quantum mechanical. Add to this the obvious fact that any macroscopic device is made up of a multitude of particles obeying quantum physics, and the problem seems even more unresolvable.

This two-domain distinction between micro and macro phenomena in the orthodox quantum measurement theory has led to enormous confusion about the character of quantum reality. As we shall see, Bohr understood the sources of this confusion. Because the assumption that physical reality is neatly divided into separate domains disguises the fact that we live in a quantum universe, it contributes to a refusal to recognize that quantum physics constitutes the most complete description of physical reality.

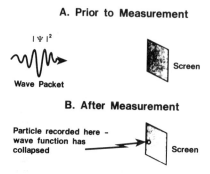

A. Prior to Measurement

$|\Psi|^2$

Wave Packet

Screen

B. After Measurement

Particle recorded here – wave function has collapsed

Screen

FIGURE 9. The square of the wave function gives the probability of finding the particle somewhere before the act of measurement. After the measurement, the wave function is said to collapse, and the particle is found at a specific location in space.

In 1932, John von Neumann developed another version of quantum measurement theory. In this version, the assumption is that both the quantum system and the measurement devices are describable in terms of what Bohr viewed as only one complementary aspect of the total reality—the wave function. In the absence of a mechanistic description of when and how the collapse of the wave function occurs, von Neumann concluded that it must occur in the consciousness of human beings. Conferring reality on only one aspect of the total reality not only results in unacceptable levels of ambiguity (if not absurdity), as Bohr said it would, but is also not consistent with experimental conditions and results in quantum physics.

The Two-Slit Experiment

One of the easiest ways to demonstrate that wave-particle dualism is a fundamental dynamic of the life of nature is to examine the results of the famous two- slit experiment. As physicist Richard Feynman put it, "any other situation in quantum mechanics, it turns out, can always be explained by saying, 'You remember the case with the experiment with two holes? It's the same thing.' "[4] In our idealized two-slit experiment we have a source of quanta, electrons, an electron gun like that in TV sets, and a screen with two openings that are small enough to be comparable with the de Broglie wavelength of an electron. Our detector is a second screen, like a TV screen, which flashes when an electron impacts on it. The apparatus allows us to record where and when an electron hits the detector.

With both slits S_2 and S_3 open (see Fig. 10), each becomes a source of waves. The waves spread out spherically, come together, and produce interference patterns that appear as bands of light and dark on our detector. In terms of the wave picture, the dark stripes reveal where the waves have canceled each other out, and the light stripes show where they have reinforced one another. If we close one of the openings, there is a bright spot on the detector in line with the other opening. The bright spot results from electrons impacting the screen in a direct line with the electron gun and the opening like bullets. Because we see no interference patterns or wave aspect, this result can be understood by viewing electrons as particles.

Physics has recently provided us with the means of conducting this experiment with a single particle and its associated wave packet arriving one at a time. Viewing a single electron as a particle, or as a point-like something, we expect it, with both slits open, to go through one slit or the other. How could a single, defined something go through both? But if we conduct our experiment many times with both slits open, we see a build-up of the interference patterns associated with waves. Because the single particle has behaved like a wave with both slits open, it does, in fact, reveal its wave aspect, and yet we have no way of knowing which slit the supposedly particle-like electron passed through.

Suppose that we refine our experiment a bit more and attempt to determine which of the slits a particular electron passes through by putting a detector (D_2 and D_3) at each slit (S_2 and S_3). After we allow many electrons to pass through the slits, knowing from the detectors which slit each electron has passed through, we discover two bright spots in a direct line with each opening that a detector indicated the electron passed through. Because no interference patterns associated with the wave aspect are observed, this is consistent with the particle aspect of the electron. Yet the choice to measure or observe what happens at the two slits reveals only the particle aspect of the total reality, and we cannot predict which detector at which slit will fire or click. All we can know is that there is a fifty percent probability that the electron in its particle aspect will be recorded at one slit or the other.

Let us now try to manipulate this reality into revealing one aspect or the other of itself by making extremely rapid changes in our experimental apparatus. The new experiment involves the double-slit arrangement with one modification—the photographic plates at the two slits are sliced so they act like Venetian blinds. Closing the blinds at the two slits creates interference patterns associated with the wave aspect of the photon. Opening the blinds at the two slits reveals the photon's particle aspect as one or the other of the two detectors placed some distance behind the slits detects the particle aspect.

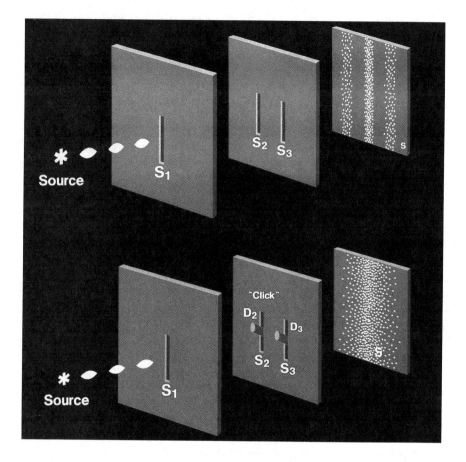

FIGURE 10. Double-slit experiment.

Now suppose that we open or close the Venetian blind-like plates at the two slits after the photon has traveled through the slits. Let us then determine if one or the other detector reveals the particle aspect in a single click or if interference patterns associated with waves are registered in the same manner as in the two-slit experiment discussed earlier. This arrangement was originally proposed in 1978 by the physicist John A. Wheeler in a thought experiment known as the "delayed-choice" experiment.

According to the predictions of Wheeler's thought experiment (see Fig. 11), when the blind is closed after the particle aspect of the photon has passed through the open slits, we should find that the screen registers the wave aspect in interference patterns and when the blinds are opened after the wave-like aspect of the photon has passed through the closed slits, we should find that the particle aspect is observed with a single click at one of the two detectors. What we are doing is determining the state of the photon with an act of observation after the photon has passed through the slits. As Wheeler put it, "we decide, after the photon has passed through the screen, whether it shall have passed through the screen."[5]

Although the delayed-choice experiment was originally merely a thought experiment, we have been able to conduct actual delayed-choice experiments with single photons. Amazingly enough, these single photons follow two paths, or one path, according to a choice made after the photon has followed one or both paths.[6] Two groups, consisting of experimental physicists at the University of Maryland and the University of Munich, found that Wheeler's predictions were borne out in the laboratory (see Fig. 12). These results indicate that wave-like or particle-like status of a photon at one point in time can be changed later in time by choosing to measure or observe one of these aspects despite the fact that the photon is traveling at the speed of light.

The results of these and other experiments show not only that the observer and the observed system cannot be separate and distinct in space; they also reveal that this distinction does not exist in time. It is as if we "caused" something to happen "after" it has already occurred. These experiments, like those testing Bell's theorem, unambiguously disclose yet another of the strange aspects of the quantum world—the past is inexorably mixed with the present and even the phenomenon of time is tied to specific experimental choices.

FIGURE 11. The delayed-choice version of the double-slit experiment of light according to Wheeler's thought experiment.

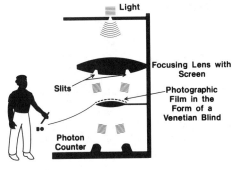

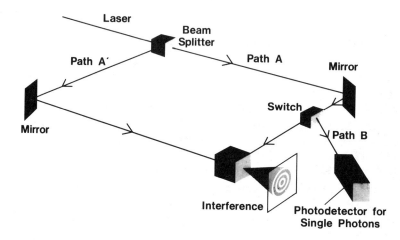

FIGURE 12. A delayed-choice experiment that has been carried out in the laboratory by groups at the University of Maryland and the University of Munich.

For the nonphysicist, it is not immediately obvious what experiments using electrons or photons can possibly say about the vast complexity of the universe in which we live. The simple answer is that what is disclosed in these experiments are general properties of all quanta, and, therefore, of fundamental aspects of everything in physical reality. Because quantum mechanical events cannot be directly perceived by the human sensorium, we are not usually aware that every aspect of physical reality "emerges" through the interaction of fields and quanta. We have only recently become fully aware of the strange properties of this reality. But if we trust the results of repeatable scientific experiments under controlled conditions, these properties are real.

Planck's Constant

The central feature of the reality disclosed in the two-slit experiments that allows us to account for the results is Planck's quantum of action. As Planck, Einstein, and Bohr showed, a change or transition on the micro level always occurs in terms of a specific chunk of energy. Nature is quite adamant about this, and there is no in-between amount of energy involved. Less than the specific chunk of energy means no transitions, and only whole chunks are involved in transitions. It is Planck's constant that weds the logically disparate constructs of wave and particle.

Let us illustrate this by returning to the double-slit experiment performed with a beam of electrons falling on a screen with two openings. Suppose we now try to predict with utmost accuracy the position and momentum of one electron.

In quantum mechanics, we find the momentum of a particle by taking Planck's constant and dividing it by the wavelength of the wave packet representing that particle.

A pure wave with a unique wavelength would have a well-defined momentum. The problem, however, is that such a wave would not be localized in any region of space and would, therefore, fill all space. Knowing the momentum precisely renders the position of the particle totally unknown.

Now suppose we try to isolate the quantum by confining it to a smaller and smaller wave packet that corresponds with the dimensions of the electron. The problem with this strategy is that as we confine the wave aspect to increasingly smaller dimensions, the number of waves increases. And because the increased number of waves of different wavelengths must be added together, this mixture of wavelengths results in a mixture of momenta. Hence, as the wave packet becomes smaller, more waves appear, and, consequently, momentum is less precise.

This is where Planck's constant, or the rule that all quantum events occur in terms of specific chunks or units of the constant, enters the picture. If Planck's constant were zero, there would be no indeterminacy, and we could predict both momentum and position with the utmost accuracy. A particle would have no wave properties and a wave no particle properties—the mathematical map and the corresponding physical landscape would be in perfect accord.

The usual value given for Planck's constant is to 6.6×10^{-27} ergs sec. Because Planck's constant is not zero, mathematical analysis reveals the following: The "spread," or uncertainty, in position times the "spread," or uncertainty, of momentum is greater than or possibly equal to the value of the constant or, more accurately, Planck's constant divided by 2π. If we choose to know momentum exactly, we know nothing about position and vice versa (see Fig. 13).

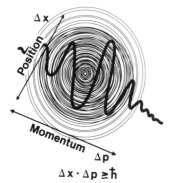

FIGURE 13. Illustration of Heisenberg's indeterminacy principle: (uncertainty in position) times (uncertainty in momentum) is at least as large as Planck's constant.

The presence of Planck's constant means that we confront in quantum physics a situation in which the mathematical theory does not allow precise prediction of, or exist in exact correspondence with, the physical reality. If nature did not insist on making changes or transitions in precise chunks of Planck's quantum of action, or in multiples of these chunks, there would be no crisis. But whether we view indeterminacy as a cancerous growth in the body of an otherwise perfect knowledge of the physical world or as the grounds for believing, in principle at least, in human freedom, one thing appears certain: it is an indelible feature of our understanding of nature.

To further demonstrate how fundamental the quantum of action is to our current understanding of the life of nature, let us attempt to do what quantum physics says we cannot do and visualize its role in the simplest of all atoms—the hydrogen atom. Imagine that you are standing at the center of the Houston Astrodome, roughly where the pitcher's mound is located. Place a grain of salt on the mound, and picture a spec of dust moving furiously around the outside of the dome in full circle around the grain of salt. This represents, roughly, the relative size of the nucleus and the distance between electron and nucleus inside the hydrogen atom when imaged in its particle aspect.

In quantum physics, however, the hydrogen atom cannot be visualized with such macro-level analogies. The orbit of the electron is not a circle in which a planet-like object moves, and each orbit is described in terms of a probability distribution for finding the electron in an average position corresponding to each orbit as opposed to an actual position. In the absence of observation or measurement, the electron could in some sense be anywhere or everywhere within the probability distribution. Also, the space between probability distributions is not empty; it is infused with energetic vibrations capable of manifesting themselves as quanta.

The energy levels manifest at certain distances because the transitions between orbits occur in terms of precise units of Planck's constant. If we attempt to observe or measure where the particle-like aspect of the electron is, as we did in the two-slit experiment, the existence of Planck's constant will always prevent us from knowing precisely all the properties of that electron which we might presume to be there in the absence of measurement. As was also the case in the two-slit experiment, our presence as observers and what we choose to measure or observe are inextricably linked to the results we get. Because all complex molecules are built up from simpler atoms, what we have said here about the hydrogen atom applies generally to all material substances.

Quantum Probabilities and Statistics

The grounds for objecting to quantum theory, the lack of a one-to-one correspondence between every element of the physical theory and the physical reality it describes, may seem justifiable and reasonable in strictly scientific terms. After all, the completeness of all previous physical theories was measured against that criterion with enormous success. Because it was this success that gave physics the reputation of being able to disclose physical reality with magnificent exactitude, perhaps a more complete quantum theory will emerge by continuing to insist on this requirement.

All indications are, however, that no future theory can circumvent quantum indeterminacy, and the success of quantum theory in coordinating our experience with nature is eloquent testimony to this conclusion. As Bohr realized, the fact that we live in a quantum universe in which the quantum of action is a given or an unavoidable reality requires a very different criterion for determining the completeness of physical theory. The new measure for a complete physical theory is that it unambiguously confirms our ability to coordinate more experience with physical reality.

If a theory does this and continues to, which is certainly the case with quantum physics, then the theory must be deemed complete. Quantum physics not only works very well; it is, in these terms, the most accurate physical theory that has ever existed. When we consider that this physics allows us to predict and measure quantities like the magnetic moment of electrons to the fifteenth decimal place, we realize that accuracy per se is not the real issue.[7] The real issue, as Bohr rightly intuited, is that this complete physical theory effectively undermines the privileged relationship in classical physics between physical theory and physical reality.

Another measure of success in physical theory is also met by quantum physics—elegance and simplicity. The quantum recipe for computing the probabilities given by the wave function is straightforward and can be successfully used by any undergraduate physics student. Take the square of the wave amplitude and compute the probability of what can be measured or observed with a certain value. Yet there is a profound difference between the recipe for calculating quantum probabilities and the recipe for calculating probabilities in classical physics.

In quantum physics, one calculates the probability of an event that can happen in alternative ways by adding the wave functions and then taking the square of the amplitude.[8] In the two-slit experiment, for example, the electron is described by one wave function if it goes through one slit and by another wave function if it goes through the other slit. To compute the probability of where the electron is going to end up on the screen, we add the two wave functions, compute the absolute value of their sum, and square it. Although the recipe in classical probability theory seems similar, it is very different. In classical physics, one would simply add the probabilities of the two alternative ways and let it go at that. The classical procedure does not work here because we are not dealing with classical atoms. In quantum physics additional terms arise when the wave functions are added, and the probability is computed in a process known as the "superposition principle."

The superposition principle can be illustrated with an analogy from simple mathematics. Add two numbers and take the square of their sum, as opposed to just adding the squares of the two numbers. Obviously, $(2+3)^2$ is not equal to $2^2 + 3^2$. The former is 25, and the latter is 13. In the language of quantum probability theory,

$$| \psi_1 + \psi_2 |^2 \neq | \psi_1 |^2 + | \psi_2 |^2$$

where ψ_1 and ψ_2 are the individual wave functions. On the left-hand side, the superposition principle results in extra terms that cannot be found on the right-hand side. The left-hand side of the preceding relation is how a quantum physicist would compute probabilities and the right-hand side is the classical analog. In quantum theory, the right-hand side is realized when we know, for example, which slit the electron went through. Heisenberg was among the first to compute what would happen in an instance like this. The extra superposition terms contained in the left-hand side of the preceding relation would not be there, and the peculiar wave-like interference pattern would disappear. The observed pattern on the final screen would, therefore, be what one would expect if electrons were behaving like bullets, and the final probability would be the sum of the individual probabilities.[9] But when we "know" which slit the electron went through, this interaction with the system causes the interference pattern to disappear.

To give a full account of quantum recipes for computing probabilities, one has to examine what would happen in events that are compound. Compound events are "events that can be broken down into a series of steps, or events that consist of a number of things happening independently."[10] The recipe here calls for multiplying the individual wave functions and then following the usual quantum recipe of taking the square of the amplitude.

The quantum recipe is $\mid \psi_1 \cdot \psi_2 \mid^2$ and, in this case, it would be exactly the same if we multiplied the individual probabilities, as one would in classical theory. Thus the recipes of computing results in quantum theory and classical physics can be totally different. The quantum superposition effects are completely nonclassical, and there is no mathematical justification per se why the quantum recipes work. What justifies the use of quantum probability theory is the same thing that justifies the use of quantum physics: It has allowed us in countless experiments to vastly extend our ability to coordinate experience with nature.

The view of probability in the nineteenth century was greatly conditioned and reinforced by classical assumptions about the relationship between physical theory and physical reality. In this century, physicists developed sophisticated statistics to deal with large ensembles of particles before the actual character of these particles was understood. Classical statistics, developed primarily by James C. Maxwell and Ludwig Boltzmann, was used to account for the behavior of molecules in a gas and to predict the average speed of a gas molecule in terms of the temperature of the gas.

The presumption was that the statistical averages were workable approximations that subsequent physical theories, or better experimental techniques, would disclose with exact precision and certainty. Because nothing was known about quantum systems and quantum indeterminacy is small when dealing with macro-level effects, this presumption was reasonable. We now know, however, that there are quantum mechanical effects in the behavior of gases and that the choice to ignore them is merely a matter of convenience in getting workable or practical results. It is, therefore, no longer possible to assume that the statistical averages are merely higher-level approximations for a more exact description.

The Schrödinger Cat Paradox

Perhaps the best known defense of the classical conception of the relationship between physical theory and physical reality took the form of a thought experiment involving a cat, and this cat, like the fabulous beast invented by Lewis Carroll, appears to have become quite famous. The thought experiment, proposed by Schrödinger in 1935, is designed to parody some perceived limitations in quantum physics and, like many parodies in literature, the underlying intent was quite serious.

In the orthodox interpretation of quantum theory, the Copenhagen Interpretation, the act of measurement plays a central role. This interpretation stipulates that before the act of measurement, one cannot know which of the many possibilities implied by the wave function will materialize. Schrödinger, the father of wave mechanics, was a believer, along with Einstein, in the one-to-one correspondence between every element of the physical theory and physical reality. The intent of the thought experiment was to argue indirectly that mathematically real properties are real even in the absence of observation.

In this Rube Goldberg-like thought experiment, we are asked to first imagine that Schrödinger's cat is a collection or ensemble of wave functions that correspond with the individual quantum particles that constitute the cat. In other words, the "reality" of the cat is identified with a multitude of wave functions. The cat is first placed inside a sealed box that can release poisonous gas. The release of the gas is determined by the radioactive decay of an atom or the passage of a photon through a half-silvered mirror. Schrödinger chose to have the gas released in this way because either trigger is quantum mechanical and, therefore, indeterminate or random.

The parody of the role of the observer in orthodox quantum measurement theory takes the form of a question. Because the observer standing outside the box does not know when the gas is released, or if the cat is alive or dead, the question is, "What is happening inside the box in the absence of observation?" Although the thought experiment may seem ludicrous, the principle at issue for Schrödinger and Einstein was very serious indeed. The experiment suggests that the cat must be both alive and dead prior to the act of observation because both possibilities remain in the isolated system in the absence of observation. Thus Schrödinger is suggesting, in the effort to point up the absurdity of any alternate view, that a mathematically real property exists in the physical reality whether we observe it or not (see Fig. 14).

The essential paradox Schrödinger seeks to amplify here has been nicely described by Abner Shimony:

FIGURE 14.
Schrödinger's cat-in-box
thought experiment:
There is a 50% prob-
ability at any time that
the cat is dead or alive.

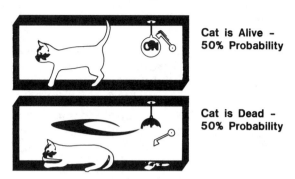

Cat is Alive –
50% Probability

Cat is Dead –
50% Probability

> There would be nothing paradoxical in this state of affairs
> if the passage of the photon through the mirror were ob-
> jectively definite but merely unknown prior to observation.
> The passage of the photon is, however, objectively indefi-
> nite. Hence the breaking of the bottle is objectively indefi-
> nite, and so is the aliveness of the cat. In other words, the
> cat is suspended between life and death until it is ob-
> served.[11]

One might be able to dismiss the paradoxical nature of this conclusion if it were supported merely by a thought experiment. But here,
as in the delayed-choice thought experiment of Wheeler, physicists
have developed actual experiments to test the paradox. Groups at the
IBM Thomas J. Watson Research Center, the AT&T Bell Laboratories, the University of California at Berkeley, and the State University of New York at Stony Brook have carried out experiments that
attempt to confirm Schrödinger's cat paradox. These experiments are
based on calculations done by Anthony J. Leggett and Sudip Chak-
ravarty that involve the quantum "tunneling" effect.

Quantum tunneling involves the penetration of an energy barrier
and is completely forbidden in classical physics. It accounts, among
other things, for the radioactive decay of nuclei and for nuclear reactions. Quantum tunneling in these experiments takes place only if a
physical quantity, a magnetic field in a superconducting ring, is in-
definite or in suspended animation. In analogy to the cat being both
dead and alive, the magnetic flux does not have one or the other of
the two possible values. What is important to realize here is that the
magnetic field in this experiment is, like the cat, a macroscopic quantity. This is what makes the analogy of the superconducting ring with
Schrödinger's cat valid, and it allows us to draw experimentally valid
conclusions about the role of the observer as it is viewed in orthodox
quantum measurement theory. In these experiments the magnetic

fields, or cats, appear to exist in two states prior to measurement or observation.

In a more recent version of this experiment, physicist Christopher Monroe and his colleagues at the National Institute of Standards and Technology succeeded in creating a superposition state in an experiment using a single beryllium atom. In this experiment, the beryllium atom was made to vibrate in such a way that a dual presence is created. The one atom, for a brief period, appears to exist in two distinct states as if two atoms existed. Here again, one cat appears to be in two "cat states" prior to observation or measurement.

But while the superconducting rings and the beryllium superposition are "macroscopic systems," the state of these systems cannot be determined until a measurement takes place, and the systems cannot be said to have a definite value before the act of observation. The state of the system is dependent on the act of observation, and its otherwise mathematically real possibilities, as given by the Schrödinger wave equation, collapse on the act of observation. If we assign a real value to the wave function in the absence of observation, then all possibilities in these macroscopic systems may seem to exist whether or not we observe them. If the superimposed states of the systems actually exist before measurement, perhaps the systems are suspended between these realities in analogy with Schrödinger's alive and dead cat.

A more careful analysis reveals, however, that this seeming paradox has nothing to do with alive or dead cats. This distorted view arises only if we insist that a real or objective description of physical reality must feature a one-to-one correspondence between the physical theory and this reality. If, however, we view this situation in terms of the actual conditions and results of quantum mechanical experiments, which Bohr's Copenhagen Interpretation requires, there is no such paradox. The state of these systems becomes real or actual when a measurement occurs, and we cannot assume the reality of potential states in the absence of measurement.

Quantum Field Theory

Contemporary physics is built on quantum mechanics extended and refined into quantum field theory. When Paul A. M. Dirac combined special relativity with quantum mechanics in 1928, the result was a relativistic quantum theory. This theory predicted the existence of positively charged electrons, termed "positrons," the anti-particles of regular electrons. The jewel of modern quantum field theory, quantum electrodynamics (QED), was developed much later. It accounts for interactions of not just electrons and positrons, but of other charged particles as well. QED is a quantum field theory of electromagnetic interactions in which electromagnetic interactions are mediated by photons. It was fully developed in the 1940s, and one of its principal architects was Richard Feynman.

All of the progress made in quantum physics indicates that the concepts of fields and their associated quanta are fundamental to our understanding of the character of physical reality. Yet these concepts, like that of four-dimensional space-time in relativity theory, are totally alien to everyday visualizable experience. Let us again, however, attempt what quantum physics deems impossible and try to visualize this unvisualizable reality.

First imagine that the universe runs like a 3-D movie. What we can detect or measure in this movie are quanta, or particle-like entities. These quanta are associated with infinitely small vibrations in what can be pictured as a grid-lattice filling three-dimensional space. Potential vibrations at any point in a field are capable of producing quanta that can move about in space and interact. There are increasingly higher energies in smaller regions in space. It is the exchange of these quanta, the carriers of the field interactions, that allows the cosmic 3-D movie to emerge and evolve in time. The projectors in this movie are the four known field interactions: strong, electromagnetic, weak, and gravitational.

In quantum field theory, particles are not acted on, as classical physics supposed, by "forces"; they "interact" with each other through the exchange of other particles. The laboratories that have provided experimental evidence confirming and refining the predictions of quantum field theory use high-energy particle accelerators. Such devices have been described as the modern equivalent of cathedrals built in the twelfth and thirteenth centuries and with good reason. They are costly and magnificent artifacts testifying to our fascination with the beauty and wonder of the universe.

The main feature of these accelerators is a large hollow ring within which electrons or protons are accelerated to great speeds and made to interact with other particles. The accelerators are not, however, "atom smashers" that "break up" matter into smaller or more basic components. The effect of the collisions is rather "transformations" in which enormous energy briefly bursts open the world of fields. These transformations provide a backward look into the high-energy regime that dominated the early life of the cosmos. As we engineer higher energy in the accelerators simulating conditions in an earlier, much hotter universe, something remarkable happens—the fields begin to blend or to transform into more unified fields. Given enough energy, which we cannot hope to produce in particle accelerators, we would be able to disclose conditions near the point of origin of the cosmos, where all the fields were unified in one fundamental field.

The general rule in physics that applies here is that an increase in energy correlates with an increase in symmetry, or in new patterns of interactions disclosing fewer contrasting elements. The expectation is that the ultimate symmetry in the cosmos at the origin would reveal no contrasts or differences in an unimaginable oneness in which no thing, or nothing, exists to be observed or measured. It would be equivalent to what mathematicians call an empty set. But even if the superconducting supercollider with its fifty-three-mile-long tunnel had been built in Texas at a cost of $4.4 billion, the energies produced would not have been sufficient to simulate conditions in the unified field. To reach energies prevalent at the beginning of the universe, we would need to build an accelerator a light year in length.

We used the analogy of the 3-D movie partially to illustrate that our normative seeing is in three dimensions, as opposed to the four-dimensional reality of space-time that the theory of relativity and quantum field theory presume. The metaphor is also useful for our purposes because 3-D movies require that we put on glasses to view them. The putting on of glasses to view a 3-D movie can be likened to acts of making observations or measurements of micro-level events in the cosmic movie. The action in the movie that we might presume to be there in the absence of measurement, or before putting on the glasses, is not the same as that which we actually observe in physical experiments. In this cosmic movie, we are confronted with two logically antithetical aspects of one complete drama, and the price of admission is that we cannot perceive or measure both simultaneously.

The central feature of quantum field theory, says Steven Weinberg, is that "the essential reality is a set of fields subject to the rules of special relativity and quantum mechanics; all else is derived as a consequence of the quantum dynamics of those fields."[12] The quantization of fields is essentially an exercise in which we use complex mathematical models to analyze the field in terms of its associated quanta, and material reality as we know it in quantum field theory is constituted by the transformation and organization of fields and their associated quanta (see Table 1). Hence this reality reveals a fundamental complementarity between particles that are localizable in space-time and fields that are not. In modern quantum field theory, all matter is composed of six strongly interacting quarks and six weakly interacting leptons. The six quarks are called up, down, charmed, strange, top and bottom, and have different rest masses and functional charges. The up and down quarks combine through the exchange of gluons to form protons and neutrons.

Table 1. Four Elementary Interactions Found in Nature

Interaction	Physical phenomena	Range[a]	Relative strength	Radiation quanta[b]	Matter quanta[c]
Strong	Nuclear force	10^{-13} cm	1	Gluons	Quarks[d]
Electromagnetic	Atomic forces, optics, electricity	∞	10^{-2}	Photon	Quarks, leptons (i.e., electrons, muons)
Weak	Radioactivity, nuclear reactions in stars	10^{-16} cm	10^{-10}	$W^{\pm}$,Z	Quarks, leptons, neutrinos
Gravitational	Planetary orbits, binary stars, galaxies, clusters of galaxies, black holes, etc.	∞	10^{-38}	Graviton	Everything

[a]10^{-13} cm is about the size of a nucleus.
[b]These are the quanta responsible for the transfer of the interaction.
[c]These particles interact with each other under the particular interaction.
[d]Quarks are the constituent particles of neutrons and protons and other heavy particles.

In quantum field theory, potential vibrations at each point in the four fields are capable of manifesting themselves in their complementary aspect as individual particles, and the interactions of the fields result from the exchange of quanta that are carriers of the fields. The carriers of the field, known as messenger quanta, are the graviton for gravity, the photon for electromagnetism, the intermediate bosons for the weak force, and the "colored" gluons for the strong-binding force. If we could recreate the energies that were present in the first trillionths of trillionths of a second in the life of the universe, these four fields would, according to quantum field theory, become one fundamental field.

The movement toward a unified theory has evolved progressively from supersymmetry to supergravity to string theory. In string theory the one-dimensional trajectories of particles, illustrated here in the Feynman diagrams (see Fig. 15), is replaced by the two-dimensional orbits of a string. In addition to introducing the extra dimension, represented by the very small diameter of the string, string theory also features another small but nonzero constant that is analogous to Planck's quantum of action. Because the value of the constant is very small, it can be generally ignored except at extremely small dimensions. But because the constant, like Planck's constant, is not zero, this results in departures from ordinary quantum field theory in very small dimensions.

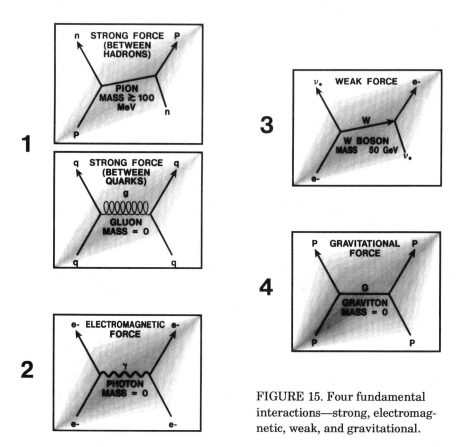

FIGURE 15. Four fundamental
interactions—strong, electromagnetic, weak, and gravitational.

Part of what makes string theory attractive is that it eliminates,
or "transforms away," the inherent infinities found in the quantum
theory of gravity. If the predictions of this theory are proved valid in
repeatable experiments under controlled conditions, it could allow
gravity to be unified with the other three fundamental interactions.
However, even if string theory leads to this grand unification, it will
not alter our understanding of wave-particle duality. Although the
success of the theory would reinforce our view of the universe as a
unified dynamic process, it applies to very small dimensions and does
not affect our view of wave-particle duality.

Although we do not know where the future progress of physics will
lead, one thing seems certain. This progress, like that made in the
rest of modern physics, will continue to disclose a profound new relationship between part and whole that is completely nonclassical.
Physicists, in general, have not welcomed this new relationship primarily because it unambiguously suggests that the classical concep-

tion of the ability of physical theory to disclose the whole as a sum of its parts, or to see reality-in-itself, can no longer be held as valid. What Bell's theorem and the experiments testing that theorem make clear is that these classical assumptions are no longer valid. The questions Bell posed in his theorem are those that were left unresolved in the twenty-three-year-long debate between Einstein and Bohr. In an effort to better explain just how important these questions were, we will revisit that famous debate in the next chapter.

3
Confronting a New Fact of Nature: Bell's Theorem and the Aspect and Gisin Experiments

> We may safely say that nonseparability is now one of the most certain general concepts in physics.
>
> *Bernard d'Espagnat*

The origins of the thought experiment that became the basis for the actual experiments testing the predictions of Bell's theorem can be traced to central issues in a debate between Bohr and Einstein. This debate began at the fifth Solvay Congress in 1927 and continued intermittently until Einstein's death in 1955. The argument took the form of thought experiments in which Einstein would try to demonstrate that it was theoretically possible to measure, or at least determine precise values for, two complementary constructs in quantum physics, like position and momentum, simultaneously. Bohr would then respond with a careful analysis of the conditions and results in Einstein's thought experiments and demonstrate that there were fundamental ambiguities he had failed to resolve. Although both men would have despised the use of the term, Bohr was the "winner" on all counts. Eventually, the dialog revolved around the issue of "realism," and it is this issue that Einstein felt would decide the correctness of quantum theory.

One of the early thought experiments proposed by Einstein, the so-called "clock in the box" experiment, illustrates how each stage of the debate typically played itself out. Suppose, said Einstein, we have a box that has a hole in one wall, and that this hole is covered by a shutter that can be opened and closed by the action of a clock inside the box. Also assume that the box contains radiation or photons of light and the clock opens the shutter at some precise time and allows one photon, or quantum of light, to escape before it closes. Suppose, Einstein continued, we then weigh the box before the photon is released, wait for the photon to escape at the precise predetermined time, and then weigh it again. Because mass is equivalent to energy, the difference in the two weights will allow us, he said, to determine the energy of the photon that escaped. We already know the exact time the photon escaped, so we can then, argued Einstein, know both the exact energy of the photon and the exact time it escaped (see Fig. 16). He concludes that both of these complementary aspects of the system can be known and the uncertainty principle is, therefore, refuted.

Focused, as always, on the conditions and results of experiments, Bohr showed why this procedure cannot produce the predicted result. He first noted that because the weighted box is suspended by a spring in the gravitational field of Earth, the rate at which the clock runs, as Einstein himself had demonstrated in the general theory of relativity, is dependent on its position in the gravitational field. Bohr then pointed out that as the photon escapes, the change in weight and the recoil from the escaping photon would cause the spring to contract and therefore alter the position of both box and clock. Because the positions of both change, there is some uncertainty regarding this position in the gravitational field, and therefore some uncertainty in the rate at which the clock runs.

Suppose, Einstein replied, we attempt to restore the original situation by adding a small weight to the box that would stretch the spring back to its original position and then measure the extra weight to determine the energy of the escaping photon. This strategy will not work, said Bohr, because we cannot reduce the uncertainty beyond the limits allowed by the uncertainty principle.

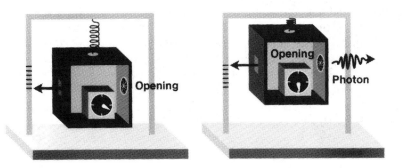

FIGURE 16. The clock-in-the-box experiment, a thought experiment devised by Einstein to refute the uncertainty principle.

Einstein was enormously persistent in his efforts to disprove the uncertainty principle and, therefore, Bohr's Copenhagen Interpretation. But he was also willing to accept the inadequacy of one thought experiment after another based on Bohr's detailed replies. What both tendencies illustrate is that Einstein knew full well that he was confronting dilemmas that dwarf any narrow concerns about professional reputation or even the merits of a physical theory.[1]

The EPR Thought Experiment

After Einstein finally accepted the idea that the uncertainty or indeterminacy principle is a fact of nature, the essential point of subsequent disagreement in the Einstein-Bohr debate became whether quantum theory was a complete theory. The more substantive point of disagreement, however, involved some profound differences concerning the special character of the knowledge we call physics. It was this issue that became the central concern in the so-called EPR thought experiment that eventually led to Bell's theorem and the experiments testing the theorem.[2]

While at Princeton in 1934 and 1935, Einstein shared his concerns with Boris Podolsky and Nathan Rosen, and the thought experiment appeared in a paper published by the three physicists in 1935. The rationale for the EPR thought experiment was the same as in all the previous thought experiments devised by Einstein in the endless debate with Bohr. Quantum mechanics is "incomplete," alleged Einstein, Podolsky, and Rosen, because it does not meet the following requirement: "Every element in the physical theory must have a counterpart in the physical reality."

The EPR thought experiment involves a new kind of imaginary test for orthodox quantum measurement theory that uses experimental information about one particle to deduce complementary properties, like position and momentum, of another particle. In this thought experiment, we are asked to imagine that two photons originate from a definite quantum state and then move apart without interacting with anything else until we elect to measure or observe one of them.

The quantum rules allow us to calculate the momentum of two particles in a definite quantum state prior to separation, and the assumption in the EPR thought experiment is that the individual momentum of the two particles will be correlated after the particles separate. If, for example, two photons originate from a given quantum state, the spin of one particle will strictly correlate with that of the other paired particle. We are then asked to measure the momentum of one particle after it has moved a sufficient distance from the other to achieve a "space-like separation." As noted earlier, this is a situation where no signal traveling at the speed of light can carry information between the two paired particles in the time allowed for measurement. Assuming that the total momentum of the two particles is conserved, we should be able, argued Einstein and his colleagues, to calculate the momentum of the paired particle that was not measured or observed based on measurement or observation of the other paired particle.

Because measurement of the momentum of one particle invokes the quantum measurement problem, Einstein conceded that we cannot know the precise position of this particle. In spite of this limitation, however, he assumed that measurement of the momentum of the particle we actually measured would not disturb the momentum of the space-like separated particle, which could be as far away from the first as one likes. Because we can calculate the momentum of the particle that was not measured and know the position of the particle that was measured, this should allow us, claimed Einstein and his colleagues, to deduce both momentum and position of the particle that was not measured. And this, they argued, would circumvent the rules of observation in quantum physics.

The point was that if we can deduce both position and momentum for a single particle in apparent violation of the indeterminacy principle, it is still possible to assume a one-to-one correspondence between every aspect of the physical theory and the physical reality. The paper concludes that the orthodox Copenhagen Interpretation "makes the reality of [position and momentum in the second system] depend upon the process of measurement carried out on the first system which does not disturb the second system in any way. No reasonable definition of reality could be expected to permit this."[3]

Bohr countered that a measurement by proxy does not count, and that one cannot attribute the reality of both position and momentum to a single particle unless you measure that particle. What would prove most important about the EPR thought experiment, however, is that it featured another fundamental classical assumption that physicists regarded at the time as an incontrovertible truth—the principle of local causes—which states that a physical event cannot simultaneously influence another event without direct mediation, such as the sending of a signal. In the EPR experiment, this means that a measurement of one particle cannot simultaneously affect the measurement of the second particle in a space-like separated region. One would have to assume, if the principle of local causes is valid, that a signal can travel faster than light for such an influence to occur, and this would force us to abandon the theory of relativity and virtually all of modern physics.

Einstein realized, of course, that the quantum formalism indicates that there should be correlations between particles like those in the EPR thought experiment regardless of the distance between the two particles or the magnitude of the space-like separation. The intent in the EPR thought experiment was, therefore, to make the following argument: Because the correlations predicted by quantum physics could not possibly occur under the experimental conditions described in the EPR experiment, this should allow us to conclude that quantum theory is incomplete and poses no challenges to the classical view of correspondence between physical theory and physical reality.

What was needed to finally settle these matters were actual experiments to test the assumptions. John Bell of the Centre for European Nuclear Research conceived of a way to accomplish this in 1964. Bell deduced, mathematically, the most general relationships between two particles, like those in the EPR experiment, and showed that certain kinds of measurement could distinguish between the positions of Einstein and Bohr. One set of experimental results would prove quantum theory complete and Bohr correct, and another set would prove quantum theory incomplete and Einstein correct.

The mathematical statement derived by Bell in his theorem is known as Bell's inequality, and it is predicated on two major assumptions in local realistic theories—locality and realism. Locality assumes that signals or energy transfers between space-like separated regions cannot occur at speeds greater than light. And realism assumes that physical reality exists independently of the observer and that the state of this reality is not dependent on acts of observation or measurement. Because the formalism of quantum physics indicates that neither assumption may be valid, the experiments testing these assumptions would resolve fundamental issues in the Einstein-Bohr debate. If these experiments revealed that Bell's inequality was violated, the fundamental issues in this debate would be resolved in favor of Bohr. The important point is that the issue could now be submitted to the court of last resort—repeatable scientific experiments under controlled conditions.

While most of the experiments testing Bell's theory involve the polarization of photons, perhaps the best way to describe what occurs in these experiments is to first talk about the spin of electrons. Assuming that paired electrons originate in a single quantum state, like that featured in the EPR experiment, they must have equal and opposite spin as they move in opposite directions from this source. But because the spin of each paired electron is quantized and obeys the uncertainty principle, not all components of the spin of a single electron can be measured simultaneously any more than position and momentum can be measured simultaneously.

A measurement of the spin of an electron on one path or the other will, therefore, yield the result "up" fifty percent of the time and "down" fifty percent of the time, and we cannot predict with any certainty what the result will be in any given measurement. When viewed in isolation, the spin of each of the paired electrons will show a random fluctuation pattern that would confuse attempts to know in advance the spin of the other. But because we also know that each of the two paired particles has equal and opposite spin, the random spins in one particle should match precisely, or correlate with, those of the other when we conduct the experiment many times and view both particles together rather than in isolation.

What we have said here about the relationship between spin states in paired electrons also applies to polarization states of paired photons. Polarization defines a direction in space associated with the wave aspect of the massless photon. The polarization of a photon, like the spin of an electron, also has a "yes" or "no" property that obeys the indeterminacy principle, and the relationship between these properties in paired photons is the same at that between paired electrons. Polarization of paired photons, like those in experiments testing Bell's theory, are equal and opposite, and the random polarization of one paired photon should precisely match or correlate with the other if the experiment is run a sufficient number of times.

A More Detailed Account of Experiments Testing Bell's Theorem

With the complementary nature of polarization in mind, the results of the experiments testing Bell's theorem can be illustrated with a simple two-photon system that uses a crystal similar to a polarizing film as a transmission device (see Fig. 17).[4] Such a crystal splits a beam of light that falls on it into one beam that is polarized linearly along the axis, or parallel to the axis, of the crystal and another beam polarized perpendicularly to the axis of the crystal. Detectors record the path of each photon correlating with either the parallel or perpendicular polarization.

Quantum theory predicts the probability of each possible experimental outcome when the photon is polarized along the optical axis of the crystal; the probability that it will pass through the crystal and be recorded along that channel is 1. If a photon is polarized perpendicular to the optical axis of the crystal, the probability of that photon passing through the crystal and being recorded along the same channel is 0. Quantum theory also predicts that if the photon is polarized linearly at some angle between 0 and 90 degrees to the transmission axis, the probability of that photon passing through the crystal is a number between 1 and 0.

Now suppose that, as in the original EPR argument, two photons originate from a single quantum state and propagate in opposite directions (see Fig. 18). In one quantum state, the overall beam by itself appears completely unpolarized, but the polarization of each photon is perfectly correlated with its partner. In other words, the total polarization of the two-photon system is such that the two individual polarizations would always have to be along the same direction in space. One possible state is when both photons are polarized along a given direction in space where the optical axis is pointing. We denote this by A, which stands for "parallel to the transmission axis." The other possible quantum state is when they are both polarized along a direction "perpendicular to the first transmission axis." We denote this second quantum state by the letter E.

The quantum superposition principle also allows the formation of a quantum state that contains equal amounts of the parallel polarized state and the perpendicular polarized state. If we insert crystals in the paths of the photons with both transmission axes straight up, this will result in both photons being in state A or state E. In other words, there is a probability of one-half, or fifty percent, that both photons will pass through channel A and a probability of one-half that both will pass through channel E. In this case we have strict correlation in the outcomes of the experiments involving the two photons.

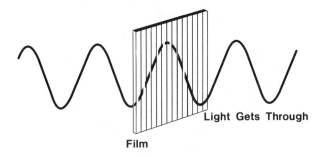

FIGURE 17. Illustration of the polarization of light measured with a piece of polarized film. Light gets through if it is polarized along the transmission axis of the film.

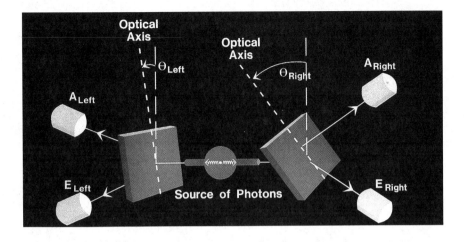

FIGURE 18. A simplified version of an experiment testing Bell's inequality.

Denote the photon that flies to the left the "left" photon and the other the "right" photon. Two typical synchronized sequences of measurements of polarization, where A stands for the photon polarized along the axis or parallel to the optical axis and E for the photon polarized perpendicular to the axis of the crystal, would then look like this:

$$\text{Left: } A\,E\,A\,E\,A\,A\,E\,A\,E\,E\,A\,A\,A$$

$$\text{Right: } A\,E\,A\,E\,A\,A\,E\,A\,E\,E\,E\,A\,A\,A \tag{1}$$

Because the actual orientation in space of the optical axis is immaterial, it does not matter to which direction in space the two optical axes point. As long as both are parallel, we can change the orientation of the axes and the records would still look similar to that shown in (1). One can keep track of the two optical axes of the crystals by constructing dials that read a direction in space like the hand of a clock. If both optical axes are at any angle, say, along the 12:00 direction, 2:00 direction, 7:00 direction, and so on, the measurement records in all these cases will be similar to (1).

The word "similar" is important here because any finite number of measurements will not necessarily look identical to (1). If, however, many measurements are made, quantum probability predicts that fifty percent of the time both left and right will record an A polarization, and that fifty percent of the time they will both record an E polarization. Given a sufficient number of measurements, we should discover that photons are polarized along the given direction fifty percent of the time and photons are polarized perpendicularly to the given direction fifty percent of the time.

Suppose we force the optical axis of the left crystal to be, say, along the 12:00 direction, and the one on the right at 90 degrees, or the 3:00 direction. The sequences of measurement will now look like:

$$\text{Left: } A\,E\,A\,A\,E\,E\,A\,E\,A\,A\,E\,E\,A\,E\,E$$

$$\text{Right: } E\,A\,E\,E\,A\,A\,E\,A\,E\,E\,E\,A\,A\,E\,A\,A \tag{2}$$

This means we have perfect anti-correlation between the polarizations of the two photons. When the left paired photon passes through the 12:00 crystal and is recorded by the A detector, it has a polarization parallel to it. But when the right paired photon is recorded by the E detector, it has a polarization along the 3:00 direction. The opposite is true of the right photon.

Because we go from the perfect matching of the sequences (1), when both axes are along the same direction, to the perfect mismatching of the sequences (2), when one axis is perpendicular to the other, there must be intermediate orientations in the two directions where we do not find perfect matchings or perfect mismatchings. In particular, there must be an intermediate angle between the two orientations for which there are three matches out of four and one mismatch out of four. The sequence of measurements will then look like this with the mismatches underlined:

$$\text{Left: } A\,E\,\underline{E}\,\underline{E}\,\underline{A}\,A\,E\,E\,A\,\underline{E}\,A\,A\,E\,A\,A\,\underline{E}$$

$$\text{Right: } A\,E\,\underline{A}\,E\,\underline{E}\,A\,E\,E\,A\,\underline{A}\,E\,E\,A\,E\,E\,\underline{A} \tag{3}$$

Quantum theory actually says that the angle between the two orientations is 30 degrees. If the left crystal axis is along the 12:00 direction, the right axis will have to be placed along the 1:00 direction. If the left crystal axis is along the 3:00 direction, the right axis will have to be placed along the 4:00 direction, and so on.

Finally, there must be another angle between the two orientations for which there are three mismatchings out of four and one matching out of four. The sequence of measurements will then look like this, with the mismatches underlined once again:

$$\text{Left: } A \underline{E} \, \underline{E} \, \underline{E} \, A \, A \, A \underline{E} \, A \underline{E} \, A \, A \underline{E} \, A \underline{E} \, A \underline{E}$$

$$\text{(4)}$$

$$\text{Right: } A \underline{A} \, \underline{A} \underline{E} \, \underline{E} \, \underline{E} \, \underline{E} \, \underline{A} \underline{A} \, \underline{A} \underline{A} \, \underline{E} \, \underline{A} \underline{E} \, \underline{A} \, A$$

Quantum theory predicts that the angle between the two orientations is 60 degrees. If the left axis is along the 12:00 direction, the right axis would be along the 2:00 direction.

To summarize, quantum theory predicts the sequences (1), (2), (3), and (4) for the four angles between the two axes equal to 0, 90, 30, and 60 degrees, respectively. What the actual experiments carried out in the laboratory to test Bell's theorem have shown is that the predictions of quantum theory are valid and that Bell's inequality is violated in accordance with the predictions of quantum theory.

Results of Experiments Testing Bell's Theory

The results of experiments testing Bell's theorem clearly reveal that Einstein's assumption in the EPR thought experiment that correlations between paired protons over space-like separated regions could not possibly occur was wrong. The experiments show that the correlations do, in fact, hold over any distance instantly, or in "no time." Because this violates assumptions in local realistic theories, physical reality is not, as Einstein felt it should and must be, local. The experiments clearly indicate that physical reality is nonlocal.

If we can imagine that both Einstein and Bohr were somehow alive and well when the results of experiments testing Bell's theorem were published, each would realize that their famous debate had finally been resolved in Bohr's favor. Both would readily appreciate the fact that if physical reality is nonlocal, quantum indeterminacy and the quantum observation problem cannot be obviated or subverted under any experimental conditions. Realizing that this is the case, Einstein would probably have been among the first to concede that a one-to-one correspondence between physical reality and physical theory does not exist in a quantum mechanical universe. Given that this was Einstein's only final point of disagreement with Bohr's Copenhagen Interpretation, one must also imagine that he would concede that this interpretation must now be viewed as the only valid interpretation.

Other physicists, most notably David Bohm and Louis de Broglie, have sought to undermine Bohr's Copenhagen Interpretation with the assumption that the wave function does not provide a complete description of the system. If this were the case, then one could avoid the conclusion that quantum indeterminacy and probability are inescapable aspects of the quantum world and assume that all properties of a quantum system can be known in principle if not in practice. What these physicists have attempted to do is assign a complete determinacy at an unspecified subquantum level. They speculate that a number of variables that are inaccessible to the observer at both the macro and quantum levels exist on this level.

These so-called "hidden variables" would supposedly make a quantum system completely deterministic at the subquantum level. Although quantum uncertainty or indeterminacy is apparent in the quantum domain, the assumption is that determinism reigns supreme at this underlying and hidden level. This strategy allows one to assume that although quantum indeterminacy may be a property of a quantum system in practice, it need not be so in principle. It also allows one to view physical attributes of quantum systems, such as spin and polarization, as objective or "real" even in the absence of measurement, and to assume, as Einstein did, a one-to-one correspondence between every element of the physical theory and physical reality.

One large problem with these so-called "local realistic classical theories" is that they cannot be verified in experiments. Another is that they predict a totally different result for the correlations between the two photons in experiments testing Bell's theorem, and this was one of Bell's motives for deriving his theorem. The assumption that the variables are hidden or unknown will obviously not allow us to determine whether what happens at the left filter in experiments testing Bell's theorem is causally connected to what happens at the right filter, but we can test the reasonableness of hidden-variable theories here with a simple assumption.

If locality holds, or if no signal can travel faster than light, turning the right filter can change only the right sequence and turning the left filter can change only the left sequence. According to hidden-variable theories, turning the second axis from the 12:00 direction to the 1:00 direction should yield one miss out of four in the right sequence, and turning the first axis from the 12:00 direction to the 11:00 direction should yield one miss out of four in the left sequence. If we take into account the overlaps in the mismatches between the two sequences, we could conclude that the overall mismatching rate between the two sequences is two or less out of four. Local realistic theories, or hidden-variable theories, would therefore predict the following sequences of measurements:

Left: $A \underline{E} \underline{A} A E \underline{A} E E \underline{E} \underline{A} A \underline{A} \underline{A} E \underline{A} E$

(5)

Right: $A \underline{A} \underline{E} A E \underline{E} E E \underline{A} A A \underline{E} \underline{E} \underline{E} E$

It is clear when one compares (4) with (5) that "for certain angles" local realistic theories would predict records that differ significantly in their statistics from what quantum theory predicts. Bell's theorem both recognizes and states this fact. The specific way local realistic theories differ from quantum theory is given by various kinds of Bell inequalities, and it is clear that quantum theory strongly violates such inequalities for certain angles, such as 60 degrees in the example presented here.

History of Experiments Testing Bell's Theorem

The first tests of Bell's inequality were conducted at the University of California, Berkeley, and the results were reported in 1972. In the earliest tests, photons were emitted from calcium or mercury atoms that were excited into a specific energetic state by laser light. The return to the ground state from the excited state involved an electron in two transitions between an intermediate state and the ground state, and a photon is created in each transition. The two photons were produced for the transitions chosen with correlated polarizations. Using photon counters placed behind polarizing filters, the photons from the cascade were then analyzed.

In the 1970s experiments were conducted in which the photons were gamma rays produced when an electron and a positron annihilate, and the polarizations of the two photons were correlated. In the many tests that have been conducted since the 1970s, one impulse has been to eliminate any problems in the design of earlier experiments and to make the statistics as "clean" as possible. Another has been to insure that the detectors are placed far enough apart so that no signal traveling at light speed can be assumed to be accounting for the correlations.

These experiments produced results in accord with the predictions of quantum theory, and therefore they violated Bell's inequality. But it was still possible to assume that the wave function in the two-photon system was a "single wave" extending from the source to the location of the detectors and therefore that this wave carries information about the system. This assumption allows one to avoid confronting the prospect that the correlations violated locality, or occurred faster than the time required for light to carry signals between the two regions. Although the notion of a single wave function carrying information over any distance in no time from one source was "a single" wave function is a bit strange, it is at least a basis for denying the existence of nonlocality.

What was needed to dispel this notion was an experimental arrangement in which the structure of the experiment could be changed when the photons were in flight from their source. It was an arrangement of this sort that was the basis for the experiments conducted at the Institute of Optics at the University of Paris at Orsay by Aspect and his colleagues (see Fig. 19). This arrangement allowed the polarizations of the paired protons to be changed using a pseudo-random signal while they are in flight, moving toward the detectors. The results provided unequivocal evidence that the "single-wave" hypothesis is false and that Einstein's view of realism does not hold in a quantum mechanical universe. As the French physicist Bernard d'Espagnat put it in 1983, "Experiments have recently been carried out that would have forced Einstein to change his conception of nature on a point he always considered essential.... we may safely say that nonseparability is now one of the most certain general concepts in physics."[5]

In the Aspect experiments, the choice between the orientations of the polarization analyzers is made by optical switches while the photons are flying away from each other.[6] The beam can be directed toward either of two polarizing filters that measure a different direction of polarization, and each has its own photon detector behind it. The switching between the two different orientations took only 10 nanoseconds, or 10×10^{-9} sec as an automatic device generated a pseudorandom signal. Because the distance between the two filters was 13 meters, no signal traveling at the speed of light could be presumed to carry information between the filters. A light signal would take 40 nanoseconds to go from one filter to the other. This means, assuming that no signal can travel faster than light, that the choice of which orientation of polarization is measured on the right could not influence the transmission of the photon through the left filter. The results of these experiments agree with quantum mechanical predictions of strong correlations, and Bell's inequality is violated.

The recent experiments by Nicolus Gisin and his team at the University of Geneva provided even more dramatic evidence that nonlocality is a fact of nature. The Gisin experiments were designed to determine whether the strength of correlations between paired photons in space-like separated regions would weaken or diminish over significantly large distances. This explains why the distance between the detectors was extended in the Gisin experiments to 11 kilometers, or roughly seven miles.

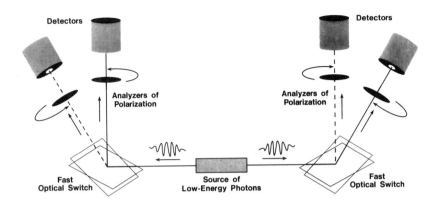

FIGURE 19. The experimental setup used by Aspect and his co-workers was designed to test the predictions of quantum theory versus the predictions of local realistic theories. There is now general agreement that the experiments testing Bell's theorem have made local realistic theories, like deterministic hidden variables, scientifically gratuitous at best.

A distance of 11 kilometers is so vast compared with distances on the realm of quanta that the experiments were essentially seeking to determine whether the correlations would weaken or diminish over any distance, no matter how arbitrarily large. If the strength of the correlations held at 11 kilometers, physicists were convinced they would also hold in an experiment where the distances between the detectors was halfway to the edge of the entire universe. If the strength of the correlations significantly weakened or diminished, physical reality would be local in the sense that nonlocality does not apply to the entire universe. This did not prove to be the case. The results of the Gisin experiments provided unequivocal evidence that the correlations between detectors located in these space-like separated regions did not weaken as the distance increased, and this obliged physicists to conclude that nonlocality or nonseparability is a global or universal dynamic of the life of the cosmos.

One of the gross misinterpretations of the results of these experiments in the popular press was that they showed that information traveled between the detectors at speeds greater than light. This was not the case, and relativity theory, along with the rule that light speed is the speed limit in the universe, was not violated. The proper way to view these correlations is that they occurred instantly, or in "no time," despite the vast distance between the detectors. The results also indicate that similar correlations would occur even if the distance between the detectors were billions of light years.

A number of articles in the popular press also claimed that the results of the Gisin experiments showed that faster-than-light communication is possible. This misunderstanding resulted from a failure to appreciate the fact there is no way to carry useful information between paired particles in this situation. The effect studied in the modern EPR-like experiments applies only to events that have a common origin in a unified quantum system, like the annihilation of a positron-electron pair, the return of an electron to its ground state, or the separation of a pair of photons from the singlet state. Because any information that originates from these sources is, in accordance with quantum theory, a result of quantum indeterminacy, the individual signals are random, and random signals cannot carry coded information or data.

The polarizations, or spins, of each photon in the Gisin experiments carry no information, and any observer of the photons transmitted along a particular axis would see only a random pattern. This pattern makes nonrandom sense only if we can compare it with the pattern observed in the other paired photon. Any information contained in the paired photons derives from the fact that the properties of the two photons exist in complementary relation, and that information is uncovered only through a comparison of the difference between the two random patterns.

Confronting a New Fact of Nature

Although the discovery that nonlocality is a fact of nature will not result in a technological revolution in the telecommunications industry, it does represent a rather startling new addition to our scientific world-view. As Henry Stapp puts it, nonlocality could be the "most profound discovery in all of science."[7] The violation of Bell's inequality also requires us to make some profound revisions in our understanding of the character of the knowledge called physics. The assumption in Einstein's thought experiment was that if we can predict with certainty in physical theory the value of a physical quantity without disturbing the system, then this element of the physical theory fully corresponds with the quantity in physical reality.

Although Bell's theorem, which is based on two particles and their associated inequalities, does not speak to this issue, one can show that this correspondence would not exist in EPR-like experiments involving three or more particles. In this situation, the violation of Bell's inequalities would be much more severe and would grow exponentially in proportion to the number of entangled particles in the original quantum state. If EPR-like experiments on three or more particles could be conducted, deterministic models based on the assumptions of locality and realism could not explain the results, and the lack of correspondence between every element in the physical theory and physical reality would be apparent in a startling new way.[8]

It is also important to realize that the Aspect and Gisin experiments reveal, as Bernard d'Espagnat pointed out, a general property of nature.[9] All particles in the history of the cosmos have interacted with other particles in the manner revealed by the Aspect experiments. Virtually everything in our immediate physical environment is made up of quanta that have been interacting with other quanta in this manner from the big bang to the present. Even the atoms in our bodies are made up of particles that were once in close proximity to the cosmic fireball, and other particles that interacted at that time in a single quantum state can be found in the most distant star. Also consider, as the physicist N. David Mermin has shown, that quantum entanglement grows exponentially with the number of particles involved in the original quantum state and that there is no theoretical limit on the number of these entangled particles.[10] If this is the case, the universe on a very basic level could be a vast web of particles that remain in contact with one another over any distance in no time in the absence of the transfer of energy or information.

This suggests, however strange or bizarre it might seem, that all of physical reality is a single quantum system that responds together to further interactions. The quanta that make up our bodies could be as much a part of this unified system as the photons propagating in opposite directions in the Aspect and Gisin experiments. Thus nonlocality, or nonseparability, in these experiments could translate into the much grander notion of nonlocality, or nonseparability, as the factual condition in the entire universe.

There is little doubt among physicists that nonlocality must now be recognized as a fact of nature, but not much has been done to explore the larger implications beyond the conclusion that Bohr's Copenhagen Interpretation of quantum mechanics must remain the "orthodox" interpretation. We will now examine the implications of this fact of nature for scientific epistemology or for our scientific worldview generally. Basic to this discussion will be our new understanding of a fundamental relationship—that between the part and the whole as has been disclosed in physical theories since the special theory in 1905. The first task is to demonstrate that the meaning of the principle of complementarity as defined by Niels Bohr has not been well understood among the community of physicists.

4
Changing the Rules: A New Epistemology of Science

I am afraid of this word Reality.

Arthur Eddington

As we have seen, the central pillar of Bohr's Copenhagen Interpretation is complementarity. The usual textbook definition of "complementarity" says that it applies to "apparently" incompatible constructs, such as waves and particles, or variables, such as position and momentum. Because one of the paired constructs or variables cannot define the situation in the quantum world in the absence of the other, both are required for a complete view of the actual physical situation. Thus a description of nature in this special case requires that the paired constructs or variables be viewed as complementary, meaning that both constitute a complete view of the situation but only one can be applied in a given situation. The textbook definition usually concludes with the passing comment that because the experimental situation determines which complementary construct or variable will be displayed, complementarity assumes that entities in the quantum world, like electrons or photons, do not have definite properties apart from our observation of them.

One reason why complementarity is dealt with in such a cursory and inadequate manner in most physics textbooks is that physicists have tended to view it as a heuristic with no physical content rather than a description of an inherent aspect of physical reality. Another is that it has been possible to assume until recently that Bohr's CI either does not apply to all of physics or can be viewed as a provisional and passing interpretation. A third reason could be that Bohr's efforts to achieve the utmost clarity often resulted in a prose so riddled with qualifications that it was difficult to determine his precise meaning. When we examine his statements in the light of recent developments in physics, however, it is not difficult to see how precise they are.

Much of the confusion about Bohr's understanding of the epistemological situation in quantum physics seems to derive from his frequent description of quantum mechanics as a "rational generalization of classical mechanics" and his requirement that the results of quantum mechanical experiments "must be expressed in classical terms."[1] When these statements are read out of context, as the physicist and philosopher of science Clifford Hooker notes, one could conclude that quantum mechanics is an "extension" of classical mechanics. This seems to make legitimate the view that our experience in the quantum domain is merely a special case in which working hypotheses and assumptions from classical mechanics must be modified while remaining fundamentally unchallenged.[2]

When we look at Bohr's statements in context, however, we discover that he viewed classical mechanics as a subset of quantum mechanics, or as an "approximation" that has a limited domain of validity. Quantum mechanics, concluded Bohr, is the complete description, and the measuring instruments in quantum mechanical experiments obey this description. Although we can safely ignore quantum mechanical effects in dealing with macro-level phenomena in most circumstances because those effects are small enough to be excluded for "practical" purposes, we cannot ignore the implications of quantum mechanics on the macro level for the obvious reason that they are there. Bohr argued that because the quantum of action is always present on the macro level, this requires "a final renunciation of the classical ideal of causality and a radical revision of our attitude toward the problem of physical reality."[3]

In classical physics quantities like position and momentum, constructs like the space-time description, and laws like conservation of energy and momentum can be simultaneously applied in a single unique circumstance. Thus the results of classical experiments are precisely those predicted in physical theory. In quantum physics, however, Bohr realized that such constructs are complementary, or mutually exclusive in accordance with the indeterminacy principle. This means, he said, that the "fundamental postulate of the quantum of action...forces us to adopt a new mode of description designated as complementary in the sense that any given application of classical concepts precludes the simultaneous use of other classical concepts which in a different connection are equally necessary for the elucidation of phenomena."[4]

Because the principle of complementarity will assume increasingly more importance in the remainder of this discussion, let us pause for a moment and consider why there has been a tendency to ignore its implications for all of physics. In dealing with the behavior of macro-level objects, the smallness of the quantum of action is such that we do not need to use quantum mechanics to get reliable results. Quantum indeterminacy in a flying tennis ball is, for example, exceedingly small, and the deterministic equations of classical physics are more than adequate for predicting how the ball will fly through the air.

The initial impact of the racket "causes" the ball to move in a particular direction with a particular speed, or momentum, and its subsequent motion in space seems utterly predictable. If we factor in all the initial macro-level conditions, the ball seems to appear precisely where we predicted it would. There is no reason to assume that our observations of the ball have had any effect whatsoever on these results, and it would seem insane to imagine that the ball might not appear precisely where it did had we chosen not to observe it. Our effort to coordinate experience with physical reality on the tennis court suggests that this reality is utterly deterministic. The same applies to the behavior of simple systems that we can manipulate in normative experience.

However, as Bohr realized, when we apply classical mechanics on the tennis court, or anywhere else in dealing with objects on the macro level, we are being subjected to a macro-level illusion. As Hooker puts it, "Bohr often emphasizes that our descriptive apparatus is dominated by the character of our visual experience and that the breakdown in the classical description of reality observed in relativistic and quantum phenomena occurs precisely because we are in these two regions moving out of the range of normal visualizable experience."[5] Although our experience with macro-level objects bears no resemblance to our experience with quantum particles, those objects come into existence as a result of interaction between fields and quanta. Over the past two decades, studies of nonlinear dynamics or chaos theory have shown that even the future of a classical system may be impossible to predict based on initial conditions. Although quantum physics and chaos theory do not rest on the same theoretical foundations, the fact that both reveal the existence of an inherent unpredictability in nature is worth noting.

Unrestricted causality could be assumed to exist in nature as long as it was possible to presume that all the initial conditions in an isolatable system could be completely defined and that every aspect of this system corresponds with every element of the physical theory that describes it. However, the quanta that make up macro-level systems cannot be said to have definite properties in the absence of observation. Between observations they can be in some sense, as Richard Feynman suggests, "anywhere they want" within the limits of the uncertainty principle.

When Bohr says that the quantum of action "forces" us to adopt a new "mode" of description, he is not suggesting, as Einstein derisively commented, that "the moon is not there when it is not being observed."[6] Bohr is simply describing a new epistemological situation that we are "forced" to accept because the quantum of action is, like light speed and the gravitational constant, a constant of nature. If this were not so, classical causality and classical determinism would remain firmly in place.

Because the quantum of action is a constant of nature, adopting a new mode of description is not, as Bohr's colleague Leon Rosenfeld notes, "something that depends on any free choice, about which we can have this or that opinion. It is a problem which is imposed upon us by Nature."[7] The situation is comparable, said Bohr, to the one we faced earlier in coming to terms with the implications of relativity theory:

> The very nature of quantum theory thus forces us to regard the space-time coordination and the claim of causality, the union of which characterizes the classical theories, as complementary but exclusive features of the description, symbolizing the idealizations of observation and definition respectively. Just as relativity theory has taught us that the convenience of distinguishing sharply between space and time rests solely on the smallness of velocities ordinarily met with compared to the speed of light, we learn from the quantum theory that the appropriateness of our visual space-time descriptions depends entirely on the small value of the quantum of action compared to the actions involved in ordinary sense perception. Indeed, in the description of atomic phenomena, the quantum postulate presents us with the task of developing a "complementary" theory the consistency of which can be judged only by weighing the possibilities of definition and observation.[8]

Just as we can safely disregard the effects of the finiteness of light speed in most applications of classical dynamics on the macro level because the speed of light is so large that relativistic effects are negligible, so can we disregard the quantum of action on the micro level because its effects are so small. However, everything we deal with on the macro level obeys the rules of relativity theory and quantum mechanics and, as chaos theory has shown, unrestricted classical determinism does not universally apply even in our dealings with macrolevel systems. Classical physics is a workable approximation that seems precise only because the largeness of the speed of light and the smallness of the quantum of action give rise to negligible effects.

The notion from classical physics that the observer and the observed system are separate and distinct is also, Bohr suggests, undermined by relativity theory before it was undermined in a slightly different way by quantum physics. Just as one cannot in relativity theory view the observer as outside the observed system because one must assign that observer particular space-time coordinates relative to the entire system, so one must view the observer in quantum physics as an integral part of the observed system. In both cases there is no "outside" perspective.

Bohr also pointed out that space and time in the new space-time continuum are complementary constructs. The complete description of this reality consists of two logically disparate constructs, and each excludes the other in application to a particular situation. Complementarity also emerges in relativity theory, notes Bohr, in the equivalence between mass and energy: mass becomes energy and energy mass in much the same way that the wave and particle aspects of quanta manifest themselves.

Realism Versus Idealism in the Quantum World

The power of Bohr's arguments derives largely from his determination to remain an uncompromising realist by insisting that all conclusions be consistent with experimental conditions and results and refusing to make metaphysical leaps. He had enormous and unfailing respect for the stern gatekeeper that has habitually stood at the door of scientific knowledge—measurement or observation under controlled and repeatable experimental conditions is necessary to confirm the validity of any scientific theory. What we know about phenomena as a result of the experiments confirming the validity of quantum physics refers exclusively, said Bohr, to the "observations and measurements obtained under specific circumstances, including an account of the whole experimental arrangement."[9]

Bohr concludes that if we view phenomena in this way we cannot conceive of the act of observation or measurement as "disturbing phenomena...or creating physical attributes of atomic objects."[10] We can assume that we "disturb" or "create" phenomena via observation or measurement only if we make the prior assumption that the atomic world is describable independent of observation and measurement. As Hooker puts it, "There is no 'disturbance' here in the classical sense of a change of properties from one as yet unknown value of some autonomously possessed physical magnitude to a distinct value of that magnitude under the causal action of the measuring instrument. Even talk of change of properties, or creation of properties, is logically out of place here because it presupposes some autonomously existing atomic world which is describable independently of our experimental investigation of it."[11] The hard lesson here from the point of view of classical epistemology is that there is no godlike perspective from which we can know physical reality "absolutely in itself." What we have instead is a mathematical formalism through which we seek to unify experimental arrangements and descriptions of results.

"The critical point," said Bohr, "is here the recognition that any attempt to analyze, in the customary way of physics, the 'individuality' of atomic processes, as conditioned by the quantum of action, will be frustrated by the unavoidable interaction between the atomic objects concerned and the measuring instruments indispensable for that purpose."[12] Although we are doing what we have always done in physics—setting up well-defined experiments and reporting well-defined results—the difference is that any systematized, definite statements about results must include us and our measuring apparatus.

Because the quantum of action is unavoidably present, a one-to-one correspondence between the categories associated with the complete theory and the quantum system can never be reflected in those results. For this reason, concludes Bohr, "radiation in free space as well as isolated material particles are abstractions, their properties being definable and observable only through their interactions with other systems."[13] When we use classical terms to describe the state of the quantum system, we simply cannot assume that the system has properties independent of the act of observation. We can make that assumption only in the absence of observation.

What is dramatically different about this new situation is that we are "forced" to recognize that our knowledge of the physical system cannot in principle be complete or total. Although in quantum mechanics we have complementary constructs that describe the entire situation, the experimental situation precludes simultaneously application of complementary aspects of the complete description. The choice of which is applied is inevitably part of the results we get. The conceptual context of our descriptions may remain classical. But we are obliged to use a new logical framework based on a new epistemological foundation to make sense of the observed results.

Complementarity and Objectivity

Before we discuss in more detail what Bohr means by complementarity, we should dispense with another large misunderstanding of his position. Some have assumed that because Bohr's analysis of the conditions for observation precludes exact correspondence between every element of the physical theory and physical reality, he is implying that this reality does not objectively exist or that we have ceased to be objective observers of this reality. These conclusions are possible only if we equate physical reality with our ability to know it in an absolute sense. Does nature become real when we, like the God of Bishop Berkeley, have absolute knowledge of its character, or does it cease to be real when we discover that we lack this knowledge? Bohr thought not:

> The notion of complementarity does in no way involve a departure from our position as objective observers of nature, but must be regarded as the logical extension of our situation as regards objective description in this field of our experience. The recognition of the interaction between the measuring tools and the physical systems under investigation has not only revealed an unsuspected limitation of the mechanical conception of nature, as characterized by attribution of separate properties to physical systems, but has forced us, in ordering our experience, to pay proper attention to the conditions of observation.[14]

In paying proper attention to the conditions of observation, we are forced to abandon the mechanistic or classical concept of causality, and, consequently, the assumption that scientific knowledge can be complete in the classical sense. But it certainly does not follow that we have ceased to be objective observers of physical reality or that we cannot affirm the existence of that reality. It is rather that the requirement to be objective has led us in our on-going dialog with nature to a new logical framework for objective scientific knowledge which Bohr labeled "complementarity."

This new logical framework, said Bohr, "points to the logical condition for description and comprehension of experience in quantum physics."[15] Although usually referred to as the "principle of complementarity," the use of the word "principle" is unfortunate because complementarity is not a principle as that word is used in physics. Rather, complementarity is a "logical framework" for the acquisition and comprehension of scientific knowledge that discloses a new relationship between physical theory and physical reality that undermines all appeals to metaphysics.

The logical conditions for description can be briefly summarized as follows: In quantum mechanics, the two conceptual components of classical causality—space-time description and energy-momentum conservation—are mutually exclusive and can only be coordinated through the limitations imposed by Heisenberg's indeterminacy principle. The more we know about position the less we know about momentum, and vice versa. "Contradiction," as Rosenfeld explained, "arises when one tries to apply both of them to the same situation, irrespective of the circumstances of the situation.... . However, if one reflects on the use of all physical concepts, one soon realizes that any such concept can be used only within a limited domain of validity."[16]

The logical framework of complementarity is useful and necessary when the following requirements are met: (1) when the theory consists of two individually complete constructs; (2) when the constructs preclude one another in a description of the unique physical situation to which they both apply; and (3) when both constitute a complete description of that situation.

Whenever we discover a situation in which complementarity clearly applies, we necessarily confront an imposing limit to our knowledge of this situation. Knowledge here can never be complete in the classical sense because we cannot simultaneously apply the mutually exclusive constructs that constitute the complete description. The list of those situations, as we will suggest later, is longer than Bohr could have imagined, and we speculate that it will become even longer with the advance of scientific knowledge.

When Bohr first suggested that we live in a quantum mechanical universe in which classical mechanics appears complete only because the effects of light speed and the quantum of action can be safely ignored in arriving at useful results, one could still argue, as Einstein did, that quantum indeterminacy would be circumvented by a more complete theory. That has not happened, and there are no suggestions in our view that it ever will happen. If quantum physics is as rock bottom in its understanding the dynamics of physical phenomena as it now appears to be, the new situation disclosed in quantum physics cannot be relegated to the "special" case of experiments in this physics. It must apply to the entire body of knowledge we call physics with consequences, as Bohr fully appreciated, that are quite imposing.

"The notion of an ultimate subject as well as conceptions of realism and idealism," wrote Bohr, "find no place in objective description as we have defined it."[17] This means that physical laws and theories do not have, as the architects of classical physics supposed, an independent existence from ourselves. They are human products with a human history useful to the extent that they help us coordinate a greater range of experience with nature. "It is wrong," said Bohr, "to think that the task of physics is to find out how nature is. Physics concerns what we can say about nature."[18]

The Need to Use Classical Concepts

Why, then, did Bohr stipulate that we must use classical descriptive categories, like space-time description and energy-momentum conservation, in our descriptions of quantum events? If classical mechanics is an approximation of the actual physical situation, it would seem to follow that classical descriptive categories are not adequate to describe this situation. If, for example, quantities like position and momentum are "abstractions" with properties that are "definable and observable only through their interactions with other systems," why should we represent these classical categories as if they were actual quantities in physical theory and experiment? Although Bohr's rationale for continued reliance on these categories is rarely discussed, it carries some formidable implications for the future of scientific thought. The rationale is based on an understanding of the manner in which scientific knowledge discloses the subjective character of human reality:

> As a matter of course, all new experience makes its appearance within the frame of our customary points of view and forms of perception. The relative prominence accorded to the various aspects of scientific inquiry depends upon the nature of the matter under investigation...occasionally...the [very] "objectivity" of physical observations becomes particularly suited to emphasize the subjective character of experience.[19]

The history of science grandly testifies to the manner in which scientific objectivity results in physical theories that must be assimilated into "customary points of view and forms of perception." As we engage in this assimilation process, the subjective character of experience is occasionally emphasized in unexpected ways. The framers of classical physics derived, like the rest of us, their customary points of view and forms of experience from macro-level visualizable experiences. Thus the descriptive apparatus of visualizable experience came to be reflected in the classical descriptive categories.

A major discontinuity appears, however, as we move from a descriptive apparatus dominated by the character of our visualizable experience to a more complete description of physical reality in relativistic and quantum physics. The actual character of physical reality in modern physics lies largely outside the range of visualizable experience. Einstein, as the following passage suggests, was also acutely aware of this discontinuity: "We have forgotten what features of the world of experience caused us to frame [prescientific] concepts, and we have great difficulty in representing the world of experience to ourselves without the spectacles of the old-established conceptual interpretation. There is the further difficulty that our language is compelled to work with words which are inseparably connected with those primitive concepts."[20]

Bohr concluded that we must use the classical descriptive categories not because there is anything sacrosanct about them but because our ability to communicate unambiguously is bounded by our experience as macro-level perceivers. On this level the effects of light speed and the quantum of action are too negligible to condition our normative conceptions of subjective reality. As the French philosopher Henri Bergson was among the first to point out, our logic is the logic of solid bodies and is derived as a result of experience on the macro level. The psychologist Jean Piaget would later provide some substantive validity to Bergson's claim in his studies of the cognitive development of children.

Those studies indicate that logical and mathematical operations result from the internalization of operations executed originally with solid bodies.[21] The logical and mathematical operations we normally internalize through our dealings with visualizable solid objects treat these objects as categorically discrete units with separate identities in space and time. There is, therefore, no suggestion that the units are inseparably interconnected on a more fundamental level or that their identities reveal a fundamental sameness on this level. Because we are not usually aware of quantum mechanical processes that underlie or inform apparently solid objects, the operations that work well in our dealings with these objects appear to be self-evident aspects of reality-in-itself. However, even the human eye is capable of registering the impact of a single photon, and the structure of everyday objects is emergent from quantum mechanical events.

Despite the fact that we live in a quantum mechanical universe, Bohr's dealings with this fact in his orthodox version of the CI have occasioned more dogged resistance from scientists than any other "orthodox" interpretation in the history of scientific thought. Einstein and Schrödinger, as we saw in the discussion of the cat-in-a-box thought experiment, were early detractors, and the list of other prominent physicists who have sought in various ways to undermine Bohr's CI is impressively long. It includes figures like de Broglie, Bohm, Vigier, Wheeler, and the author of the theorem that would effectively undermine objections to CI, John Bell. Most of the detractors are identified as holding the so-called "realist" position, as opposed to the "instrumentalist" or "idealist" position of Bohr and others.

The choice of the term "realist" is intriguing in that those who are identified as such are, like Einstein in the EPR thought experiment, forced into the position of claiming that a quantity must be called "real" within the context of physical theory even if it cannot be disclosed by observation and measurement in a single instance. To be a realist in these terms, one must abandon the eminently realistic scientific credo that experimental evidence is an absolute requirement for the validation of physical theory.

Bohr is sometimes termed an anti-realist by historians of science primarily because he concluded that complementary aspects of a quantum system, like wave and particle, cannot be regarded as mirroring or picturing the entire object system. Yet Bohr's conclusion follows from the utterly realistic fact that our interactions with this system preclude the appearance of both in particular measurement interactions. The occasional use of the term "idealist" in reference to Bohr's position is equally misleading because it properly applies to the so-called "realists" who assert the existence of an ideal system with properties that cannot be simultaneously measured. Although the term "instrumentalist" is marginally more appropriate, it carries associations with the term "pragmatism" and suggests that there is something more essential here that physics will eventually disclose. If we want to put a proper label on Bohr's position, we should purge the term realism of prescientific associations and apply it to that position. Bohr is brutally realistic in epistemological terms.

CI and the Experiments Testing Bell's Theorem

If we view the results of the experiments testing Bell's theorem in terms of Bohr's orthodox version of the CI, there is no ambiguity. The correlations between results at points A and B are in accordance with the predictions of quantum physics; thus we appear to have a complete physical theory that coordinates our experience with this reality. Because indeterminacy is implicit in this theory and the results make no sense without it, this factual condition has important consequences that cannot be ignored.

The logical framework of complementarity, premised on the scientific precept that measurement or observation is required to validate any physical theory, also requires that the conditions for observation be taken into account in the analysis of results. These conditions dictate that the two fundamental aspects of quantum reality, wave and particle, are complementary. Although both constructs are required for a complete view of the situation, the conditions for observation or measurement preclude the simultaneous application of both constructs.

If we insist that one view of the situation is the complete description in our analysis of results, we are obliged to presume that something in A "causes" something to happen in B in accordance with the "deterministic" wave function. The resultant ambiguities are described as follows by Henry Stapp: "If one accepts the usual ideas about how information propagates through space and time, then Bell's theorem shows that the macroscopic responses cannot be independent of faraway causes. The problem is not alleviated by saying that the response is determined by 'pure chance.' Bell's theorem proves precisely that the determination of the macroscopic response must be 'nonchance,' or at least to the extent of allowing some sort of dependence of this response on faraway causes."[22] Accepting the usual ideas about how information propagates through space and time means remaining attached to the classical concepts of locality and unrestricted causality. If we insist on this perspective and refuse to apply the logical framework of complementarity, the results of the Aspect and Gisin experiments are more than ambiguous—they make no sense at all.

If we approach this situation, as Bohr says we must, with an analysis of the conditions for the experiment, it is clear that we cannot even begin to understand the correlations in the absence of the assumption of indeterminacy and we cannot, therefore, confirm the results in the absence of measurement. As the philosopher of science Henry Folse has observed, this means "apart from the interactions with the detectors," the system that yields these results "exists in a single, nonanalyzable quantum state." Our experience as macro-level perceivers may entice us to picture the system in the Bell-type experiments as consisting of "spatially separated particles fleeing a common origin," but complementarity indicates that this is a distorted view of the "wholeness" of the interaction in which the quantum system is prepared and which includes the observing apparatus.[23]

This situation seems "strange," as all our experience with the quantum world seems strange, in terms of macro-level expectations. Nonlocality indicates that space-like separated points A and B in the Aspect and Gisin experiments remain correlated in the unified system. Yet we can no more explain this scientific fact in the classical sense, or in terms of macro-level visualizations, than we can explain the quantum of action in these terms.

Nonlocality, like quantum transitions, is a fact of nature understandable to us only within the limits and epistemological implications of the indeterminacy principle. Our task is to say as much as we can about them based on an entirely objective analysis of efforts to coordinate experience with them. More important, we can no longer rationalize this strangeness away by presuming that it applies only to the quantum world. Bohr was correct in his assumptions that we live in a quantum mechanical universe and that classical physics represents a higher-level approximation of the dynamics of this universe. If this is so, then the epistemological situation in the quantum realm can be extended to apply to all of physics.

As we hope to demonstrate later, alternatives to the CI are fatally flawed in two respects: They are not subject to experimental verification and, more interesting, they involve appeals to extra-scientific or metaphysical constructs. Why physicists would elect to advance theories that violate two fundamental tenets of scientific epistemology can be largely explained in terms of an ongoing attachment to seventeenth-century metaphysical dualism and the doctrine that the world is completely knowable in mathematical theory. However, because these tenets of classical epistemology are not in accord with what we know about the actual character of physical reality, we can no longer view physical theories as an ontological bridge between observer and observed system. They must be viewed rather as subjectively based human constructs useful to the extent that they help us coordinate greater ranges of experience with physical reality.

Complementarity and the Language of Mathematics

Virtually every major advance in modern physical theories describing the structure and evolution of the universe has been accompanied by the emergence of new complementarities. In special relativity (1905), mass and energy are logically disparate constructs that displace one another in any single physical situation, and yet both are required for a complete understanding of the situation. In general relativity (1915), space and time are revealed as profound complementarities that exist within the larger whole of the space-time continuum. In quantum physics, additional profound complementarities emerged in waves-particles and fields-quanta. What is most intriguing about this consistent correlation between new physical theories and profound new complementarities is that there is no suggestion that the theorists were, consciously or unconsciously, appealing to the logical framework of complementarity. Even a very deliberate appeal to complementarity does not account for the actual presence of profound new complementarities in testable physical theories.

Because Bohr was convinced that complementarity is the "logic of nature," this was part of his explanation why advances in physical theory have disclosed profound new complementary relationships in physical reality. He also flirted with the prospect that we have been able to coordinate greater ranges of experience with nature in modern physical theories because complementarity is a fundamental logical principle in the language of mathematics. That complementarities are emergent in physical theory does not in itself, of course, support the idea that complementarity is the fundamental structuring principle in our conscious constructions of reality in mathematical language. But when we examine the relationships between primary oppositions in this language, it is not difficult to make the case that the logic that best explains the character of these oppositions is complementarity.

One of the more obvious fundamental oppositions in mathematics is that between real and imaginary numbers. Imaginary numbers can all theoretically be formed from the first imaginary number i, the square root of -1. But a mathematical operation in which we take the square root of a negative number does not make logical sense within the framework of real numbers. Similarly, real numbers are represented analytically as points on an infinitely extending straight line, and there is no way to represent real and imaginary numbers on the same line. However, real and imaginary numbers constitute the complete description of this aspect of mathematics, and they can be represented using higher dimensions on the complex plane.

A similar and equally fundamental complementarity exists in the relation between zero and infinity. Although the fullness of infinity is logically antithetical to the emptiness of zero, infinity can be obtained from zero with a simple mathematical operation. The division of any number by zero is infinity, while the multiplication of any number by zero is zero. A more general but equally pervasive complementarity in mathematical language is that between analytic and synthetic modes of description.

Analysis, the breaking up of whole sets into distinct mathematical units, is logically antithetical to synthesis, or the bringing together of many units to form a mathematical whole. Analysis is the operative mode in differential calculus, where a continuous function is divided into smaller and smaller parts resulting in infinitely small differentials. The complementary mode in integral calculus involves the addition of infinitely small differentials to obtain a continuous function. One operation cannot be performed simultaneously with the other, but both constitute the complete view or analysis of a given situation.

Another argument that complementarity is the foundational logic of mathematics concerns challenges to the classical view of mathematics. During the late nineteenth century, attempts to develop a logically consistent basis for numbers and arithmetic threatened to undermine the efficacy of the classical view of correspondence decades before the advent of quantum physics. From 1878 to 1897, Georg Cantor created a theory of abstract "sets" of entities that eventually became a mathematical discipline. A "set," as he defined it, is a collection of definite and distinguishable objects in thought or perception conceived as a "whole."

Cantor attempted to prove that the process of counting and the definition of integers could be placed on a solid mathematical foundation. His method was to repeatedly place the elements in one set into one-to-one correspondence with those in another. In the case of integers, Cantor showed that each integer $(1,2,3,...,n)$ could be paired with an even integer $(2,4,6,...n)$, and, therefore, that the set of all integers was equal to the set of all even numbers.

Amazingly, Cantor discovered that some infinite sets were larger than others and that infinite sets formed a hierarchy of ever greater infinities. After this failed attempt to save the classical view of the logical foundations and internal consistency of mathematical systems, it soon became obvious that a major crack appeared in the seemingly solid foundations of number and mathematics. Meanwhile, an impressive number of mathematicians began to see that everything from functional analysis to the theory of real numbers depended on the problematic character of number itself.

Figures like Karl Wierstrass, Richard Dedekind, Gottlob Frege, and Guiseppe Peano also attempted to posit a firm logical basis for number in an effort to preserve the classical view of correspondence between physical theory and physical reality. Wierstrass developed an arithmetization of analysis, Dedekind sought to define real numbers, and Frege, Dedekind, and Peano attempted to axiomize ordinary mathematics. For a time, at least, many of these efforts seemed promising. However, in 1898 Bertrand Russell realized that Cantor's infinite sets revealed an inconsistency that lay at the foundation of the classical view of mathematical systems.

Russell began with the assumption that the concept "set" is itself a "set" and must belong to the "set of all sets." He then wondered about sets that "include" themselves as members and sets that specifically "exclude" themselves as members. For example, the set of all large sets is a large set that should include itself, but the set of all students is not a student and should not include itself. Russell then considered sets that do not include themselves as members and asked whether the set of all such sets is or is not a member of itself.

Suppose, for example, that we take a set of students and all other sets that do not include themselves as members, make a set of them, and ask, "Is that set of sets a member of itself or not?" The answer, which came to be known as Russell's paradox, is as follows: "If this set of sets is a member of itself, then it cannot by definition be a member of itself. But if it is not a member of itself, it must be a member because the larger set of these sets set should include itself."

However esoteric this might seem, this contradiction concerned the most basic propositions of logic and posed some large challenges to the internal consistency of mathematics and the classical view of correspondence. After Riemann and others demonstrated that self-consistent non-Euclidean geometries could be constructed, mathematicians realized they could demote an axiom to the status of a proposition or pose its converse to build new mathematical systems. Although Russell sought to eliminate his contradiction in this manner, the strategy did not work.

Knowing that Gottlob Frege had worked on the logical foundations of number, Russell asked him in a letter if he had noticed that "there is no class (as a totality) of those classes which, each taken as a totality, do not belong to themselves?"[24] Frege replied, "Your discovery of the contradiction caused me the greatest surprise and, I would almost say, consternation, since it has shaken the basis on which I intended to build mathematics."[25] Russell soon conceived of a way to rationalize, as opposed to resolve, the contradiction with his theory of types. The theory states that sets that are on the wrong level or that include themselves too often should not be included in the foundational statements of mathematics. This ad hoc solution did not, however, solve the problem, and Russell himself realized this.

Eventually, efforts to resuscitate belief in the ontological foundations of number and logic culminated in Kurt Gödel's famous proof. Gödel considered the attempt by Russell and Whitehead in *Principia Mathematica* to establish a logically consistent foundation for mathematics and rigorously proved that the foundation could never be completed.[26] Using whole numbers, Gödel demonstrated that one— and only one—of them, different from the other, could be assigned to each formula in the *Principia*. He then put the symbols, axioms, definitions, and theorems in *Principia* into one-to-one correspondence with the whole numbers in order to mirror the mathematical structure that produced them. This allowed Gödel to prove that no finite system of mathematics can be used to derive all true mathematical statements and therefore that no algorithm, or calculation procedure, can prove its own validity.[27]

What Gödel effectively demonstrated is that the character of mathematical systems is such that any scientific description of nature predicated on one-to-one correspondence between physical theory and physical reality is necessarily incomplete because it cannot prove itself. Gödel's incompleteness theorem and Heisenberg's indeterminacy principle are not directly related, but both clearly demonstrate that there is no basis for believing in the ontology of classical epistemology and the doctrine of positivism long before the experiments testing Bell's theorem provided a more dramatic demonstration that this was the case.

If complementarity is the foundational logic of mathematics, perhaps the epistemological crisis in the classical view of mathematics that began in the nineteenth century with Cantor and ended in the twentieth century with Gödel's demonstration that this view could no longer be viewed as valid is merely another demonstration that this is the case. One can also argue that this crisis curiously foreshadowed the discovery of nonlocality. First, consider that there are three things in nature we cannot described mathematically—infinity, true randomness, and continuity.

In quantum physics, randomness is a feature of indeterminacy and the quantum of action reveals that physical events on the most fundamental level are discontinuous. "Infinities" are not describable in quantum physics and are simply factored out as a matter of convenience. This is analogous to the way that calculus factors out or ignores "instantaneous" instants and time intervals, or entities with no duration and no length, with the use of "infinitesimals." Let us now add to this picture the prospect that the laws of physics, as some have argued, are statistical descriptions of stochastic processes that result in emergent order as a result of the vast number of such processes.

If we assume that complementarity is the logic that operates within the whole and that the existence of this whole can only be inferred, as opposed to proven, by the discovery of nonlocality, perhaps the solution to all these enigmas is much simpler than we have imagined. Assuming this whole is seamlessly interconnected due to quantum entanglement, perhaps this "infinity" exists in complementary relation to the "randomness" or indeterminacy witnessed in the process of observation or measurement. If this is the case, order in nature on the most primary level can be viewed as a continuous process that we can perceive, in analogy with wave-particle dualism, only in its complementary aspects of infinity and randomness.

Perhaps the reason the mathematical description of nature cannot describe these aspects of nature also explains why this description serves us well in describing other aspects. If the whole generates emergent continuity or order that we can observe only in its complementary aspects as infinity and randomness, it follows that the mathematical description of nature should break down at this event horizon. This could be the simple explanation why we cannot describe infinity, true randomness, and continuity—they are all imbedded processes in the seamlessly interconnected whole—and physics can only describe the interactions of emergent parts. But if complementarity is also the logic of nature at increasing levels of emergent order where parts exist in complementary relation to the whole, it follows that a mathematical description of nature in which complementarity is the foundational logic should describe the interactions of the parts very nicely.

If the logical framework of complementarity is fundamental to our constructions of reality in mathematical language, this could provide a partial answer to a large question confronted throughout this discussion: Why is there a correspondence between physical theory and physical reality or between the mind capable of conceiving and applying mathematical physics and the cosmos itself? Many physicists, as we have seen, are disturbed that we cannot answer this question in the old terms with an appeal to the metaphysical presuppositions of classical epistemology. Even the widespread acceptance of the essential unity of the cosmos disclosed in modern physics does not, in most instances, compensate for the feelings of loss associated with the demise of the old classical metaphysical view of the universe. However, as long as the quantum of action is fact, there can be, for all the reasons we have explored, no one-to-one correspondence between physical theory and physical reality.

This could mean, however, that our discovery that the quantum of action is fact has led us to a deeper, and perhaps far more satisfying, sense of correspondence between our knowledge of reality in physical theory and physical reality. Although the "whole" of physical reality is not fully disclosable in physical theory, perhaps we have been successful in coordinating greater levels of experience with its emergent parts because the fundamental logical principle in the behavior of the parts is also foundational to our symbolic constructions of reality in the mathematics of physical theory. It should follow, therefore, that the mathematical description of the interactions between parts should be more in accord with the actual behavior of events in nature. This does not, of course, allow us to conclude that this thesis has been proven in scientific terms, but it does suggest that the logic of complementarity could be the logic of nature and that the use of this logic as a heuristic could serve to better explain the character of other profound oppositions in natural process.

In the next chapter, we will make the case that profound complementarities have been disclosed in the study of relationships between parts and wholes in biological reality that are analogous to those previously disclosed in the study of the relationship between parts and wholes in physical reality. This not only suggests that complementarity is the logic of nature in biological reality as well; it could also provide a basis for better understanding how increasing levels of complexity in both physical and biological reality result from the progressive emergence of collections of parts that constitute new wholes that display properties and behavior that cannot be explained in terms of the sum of the parts.

We will also argue that Darwin's theory of evolution was premised on the classical paradigm in physics and that our current understanding of nature in the biological sciences requires that we revise some aspects of this theory. Equally important for our purposes, this understanding suggests that unrestricted determinism and purely reductionist methodologies cannot account for the emergent complexities in biological life.

5
The Logic of Nature: Complementarity and the New Biology

The vitalism-mechanism controversy was a preoccupation of Niels Bohr's father, a professor of physiology at the University of Copenhagen, and a frequent topic of discussion at the family residence. Although the terms are now archaic, the distinction between a living organism, which must interact with its environment, and a detailed scientific description of that organism, which must treat the system as isolated or isolatable, remains ambiguous. Bohr dealt with fundamental ambiguities in biology the same way that he dealt with fundamental ambiguities in quantum physics—analyze the conditions for observation required for unambiguous description and avoid appeals to extra-scientific or metaphysical constructs.[1]

Because the biological regularities of living organisms display an active and intimate engagement with their environment that is categorically different from that of inorganic matter, Bohr concluded that they represent profound oppositions. And because organic and inorganic matter are constructs that cannot be applied simultaneously in the same situation and yet both are required for a complete description of the situation, they must, he said, be viewed as complementary. Bohr took this argument to the next logical conclusion. Given that the lawful regularities displayed by organic and inorganic matter are not the same, perhaps a profound complementary relationship exists between the laws of physics and those of biology.[2] The following comment by Bohr serves to clarify the basis for this hypothesis:

> Analogies from chemical experience will not, of course, any more than the ancient comparison of life with fire, give a better explanation of living organisms than will the resemblance, often mentioned, between living organisms and such purely mechanical contrivances as clockworks. An understanding of the essential characteristics of living beings must be sought, no doubt, in their peculiar organi-

zation, in which features that may be analyzed by the usual mechanics are interwoven with typically atomistic traits in a manner having no counterpart in inorganic matter.[3]

Bohr suggested that any scientific description of the biochemical basis of a living organism must treat the organism as an isolated or isolatable part of the whole of life, like parts in a clockwork or machine. Hence an "understanding of living beings must be sought," he said, in "their peculiar organization" in which "features" that may be "analyzed by the usual mechanics," or the laws of physics, are interwoven with "traits" to constitute a whole that "has no counterpart in inorganic matter." The inference is that the laws of mathematical physics can only fully describe the inanimate because the application of these laws requires that we isolate the system in the act of making measurements. Because the biological regularities or "traits" of organic matter cannot be treated as isolated, the suggestion is that the description of organic matter in mathematical physics must break down at the event horizon at which those regularities come into existence.

Here again, Bohr seems remarkably prescient. For example, a complete description in mathematical physics of all the mechanisms of a DNA molecule would not be a complete description of organic matter for an obvious reason: The quality "life" associated with the known mechanism of DNA replication exists outside of the objectified description in the seamless web of interaction of the organism with its environment. This suggests that we must conclude, as Bohr did, that the laws of nature accounting for biological regularities, or the behaviors we associate with life, are not merely those of mathematical physics. Even if we could replicate all of the fundamental mechanisms of biological life by manipulating inorganic matter in the laboratory, this problem would remain. To prove that no laws other than those of mathematical physics are involved in this experiment, we would be obliged to create life in the absence of any interaction with an environment in which the life form sustains itself or interacts.

Although most physical scientists probably assume that the mechanism of biological life can be completely explained in accordance with the laws of mathematical physics, numerous phenomena associated with life cannot be explained in these terms. For example, the apparent compulsion of individual organisms to perpetuate their genes, selfish or not, is obviously a dynamic of biological regularities that is not apparent in an isolated system. This dynamic cannot be described in terms of the biochemical mechanisms of DNA or any other aspect of isolated organic matter. The specific evolutionary path followed by living organisms is unique and cannot be completely described based on an a priori application of the laws of physics.

Part and Whole in Darwinian Theory

In our view, Bohr was correct in assuming that a scientific analysis of parts cannot disclose the actual character of a living organism because that organism exists only in relation to the whole of biological life. What he did not anticipate, however, is that the whole that is a living organism appears to exist in some sense within the parts and that more complex life forms evolved in a process in which synergy and cooperation between parts (organisms) resulted in new wholes (more complex organisms) with emergent properties that did not exist in the collection of parts. More remarkable, this new understanding of the relationship between part and whole in biology seems analogous to that disclosed by the discovery of nonlocality in physics.

Because Darwin's understanding of the relationship between part and whole was essentially classical and mechanistic, the new understanding of this relationship is occasioning some revisions of his theory of evolution. Darwin made his theory public for the first time in a paper delivered to the Linnean Society in 1858. The paper begins, "All nature is at war, one organism with another, or with external nature."[4] In *On the Origin of Species*, Darwin is more specific about the character of this war: "There must be in every case a struggle for existence, either one individual with another of the same species, or with the individuals of distinct species, or with the physical conditions of life."[5] All of these assumptions are apparent in Darwin's definition of natural selection:

> If under changing conditions of life organic beings present individual differences in almost every part of their structure, and this cannot be disputed; if there be, owing to their geometrical rate of increase, a severe struggle for life at some age, season, or year, and this certainly cannot be disputed; then, considering the infinite complexity of the relations of all organic beings to each other and to their conditions of life, causing an infinite diversity in structure, constitution, habits, to be advantageous to them, it would be a most extraordinary fact if no variations had ever occurred useful to each being's own welfare, in the same manner as so many variations have occurred useful to man. But if the variations useful to any organic being ever do occur, assuredly individuals thus characterized will have the best chance of being preserved in the struggle for life; and from the strong principle of inheritance, they will tend to produce offspring similarly characterized. This principle of preservation, or the survival of the fittest, I have called Natural Selection.[6]

Based on the assumption that the study of variation in domestic animals and plants "afforded the best and safest clue" to understanding evolution,[7] Darwin concluded that nature could produce new species by cross-breeding and selection of traits. His explanation of the mechanism in nature that results in new species took the form of a syllogism: (1) the principle of geometric increase indicates that more individuals in each species will be produced than can survive; (2) the struggle for existence occurs as one organism competes with another; and (3) in this struggle for existence, slight variations, if they prove advantageous, will accumulate and produce new species. In analogy with the animal breeder's artificial selection of traits, Darwin termed the elimination of the disadvantaged and the promotion of the advantaged "natural selection."

In Darwin's view, the "struggle for existence" occurs between an atomized individual organism and other atomized individual organisms in the "same species," between an atomized individual organism of one species with that of a "different species," or between an atomized individual organism and the "physical conditions of life." The whole as Darwin conceived it is the collection of all atomized individual organisms, or parts, and the struggle for survival occurs "between" or "outside" the parts. Because Darwin viewed this struggle as the only limiting condition in the rate of increase of organisms, he assumed that the rate would be "geometrical" when the force of struggle between parts is weak and that the rate would decline as the force becomes stronger.

Natural selection occurs, according to Darwin, when variations "useful to each being's own welfare" or useful to the welfare of an atomized individual organism provide a survival advantage and the organism produces "offspring similarly characterized." Because the force that makes this selection operates outside the atomized parts, Darwin described the whole in terms of relations between the totality of parts. For example, the "infinite complexity of relations of all organic beings to each other and to their conditions of life" refers to relations between parts, and the "infinite diversity in structure, constitution, habits" refers to "advantageous" traits within the atomized parts. It seems clear in our view that the atomized individual organisms in Darwin's biological machine resemble classical atoms and the force that drives the interactions of the atomized parts, the "struggle for life," resembles Newton's force of universal gravity. Although Darwin parted company with classical determinism in the claim that changes, or mutations, within organisms occurred randomly, his view of the relationship between part and whole was essentially mechanistic.

Part-Whole Complementarity in Microbial Life

During the last three decades, a revolution has occurred in the life sciences that has enlarged the framework for understanding the dynamics of evolution. Fossil research on primeval microbial life, the decoding of DNA, new discoveries about the composition and function of cells, and more careful observation of the behavior of organisms in natural settings have provided a very different view of the terms for survival. In this view, the relationship between parts, or individual organisms, is often characterized by continual cooperation, strong interaction, and mutual dependence.

What is more interesting for our purposes is the prospect that the whole of biological life is, in some sense, present in all the parts. For example, the old view of evolution as a linear progression from "lower" atomized organisms to more complex atomized organisms no longer seems appropriate. The more appropriate view could be that all organisms (parts) are emergent aspects of the self-organizing process of life (whole), and that the proper way to understand the parts is to examine their embedded relations to the whole. According to Lynn Margulis and Dorian Sagan, this is particular obvious in the study of microbial life:

> It now appears that microbes—also called microorganisms, germs, bugs, protozoans, and bacteria, depending on the context—are not only the building blocks of life, but occupy and are indispensable to every known living structure on Earth today. From the paramecium to the human race, all life forms are meticulously organized, sophisticated aggregates of evolving microbial life. Far from leaving microorganisms behind on an evolutionary "ladder," we are surrounded by them and composed of them.[8]

During the first two billion years of evolution, bacteria were the sole inhabitants of Earth, and the emergence of more complex life forms is associated with networking and symbiosis. During these two billion years, prokaryotes, or organisms composed of cells with no nucleus, namely bacteria, transformed Earth's surface and atmosphere. It was the interaction of these "simple" organisms that resulted in the complex processes of fermentation, photosynthesis, oxygen breathing, and the removal of nitrogen gas from the air. Such processes would not have evolved, however, if these organisms were atomized in the Darwinian sense or if the force of interaction between parts existed only outside the parts.

In the life of bacteria, bits of genetic material within organisms are routinely and rapidly transferred to other organisms. At any given time, an individual bacterium has the use of accessory genes, often from very different strains, which can perform functions not performed by its own DNA. Some of this genetic material can be incorporated into the DNA of the bacteria and some may be passed on to other bacteria. What this picture indicates, as Margulis and Sagan put it, is that "all the world's bacteria have access to a single gene pool and hence to the adaptive mechanisms of the entire bacterial kingdom."[9]

Because the whole of this gene pool operates in some sense within the parts, the speed of recombination is much greater than that allowed by mutation alone or by random changes inside parts that alter interaction between parts. The existence of the whole within parts explains why bacteria can accommodate change on a worldwide scale in a few years. If the only mechanism at work were mutations inside organisms, millions of years would be required for bacteria to adapt to a global change in the conditions for survival. "By constantly and rapidly adapting to environmental conditions," wrote Margulis and Sagan, "the organisms of the microcosm support the entire biota, their global exchange network ultimately affecting every living plant and animal."[10]

The discovery of symbiotic alliances between organisms that become permanent is another aspect of the modern understanding of evolution that appears to challenge Darwin's view of universal struggle between atomized individual organisms. For example, the mitochondria found outside the nucleus of modern cells allow the cell to use oxygen and exist in an oxygen-rich environment. Although mitochondria perform integral and essential functions in the life of the cell, they have their own genes composed of DNA, reproduce by simple division, and do so at times different from the rest of the cell.

The most reasonable explanation for this extraordinary alliance between mitochondria and the rest of the cell is that oxygen-breathing bacteria in primeval seas combined with other organisms. These ancestors of modern mitochondria provided waste disposal and oxygen-derived energy in exchange for food and shelter and evolved via symbiosis into more complex forms of oxygen- breathing life. Because the whole of these organisms was larger than the sum of their symbiotic parts, this allowed for life functions that could not be performed by the mere collection of parts, and the existence of the whole within the parts coordinates metabolic functions and overall organization.[11]

Part-Whole Complementarities in Complex Living Systems

The more complex organisms that evolved from this symbiotic union are sometimes referred to in biology texts as "factories" or "machines," but a machine, as Darwin's model for the relationship part and whole suggests, is a unity of order and not of substance, and the order that exists in a machine is external to the parts. As the biologist Paul Weiss pointed out, however, the part-whole relationship that exists within and between cells in complex life forms is not that of a machine:

> In contrast to a machine, the cell interior is heaving and churning all the time; the positions of the granules or other details in the picture, therefore, denote just momentary way stations, and the different shapes of sacs or tubules signify only the degree of their filling at the moment. The only thing that remains predictable amidst the erratic stirring of the molecular population of the cytoplasm and its substructures is the overall pattern of dynamics which keeps the component activities in definable bounds and orderly restraints. These bounds again are not to be viewed as mechanical fixed structures, but as "boundary conditions" set by the dynamics of the system as a whole."[12]

The whole within the part that sets the boundary conditions of cells is DNA, and a complete strand of the master molecule of life exists in the nucleus of each cell. DNA evolved in an unbroken sequence from the earliest life forms, and the evolution of even the most complex life forms cannot be separated from the co-evolution of microbial ancestors. DNA in the average cell codes for the production of about 2000 different enzymes, and each of these enzymes catalyzes one particular chemical reaction. The boundary conditions within each cell resonate with the boundary conditions of all other cells and maintain the integrity and uniqueness of whole organisms.

Artifacts, or machines, are, in contrast, constructed from without, and the whole is simply the assemblage of all parts. Parts of machines can be separated and reassembled, and the machine will run normally, but separation of parts from the whole in a living organism results in inevitable death. "Living processes and living organisms," wrote biologist J. Shaxel, "simply do not exist save as parts of single whole organisms."[13] Hence we must conclude, as Bertalanffy does, that "mechanistic modes of explanation are in principle unsuitable for dealing with certain features of the organic; and it is just these fea-tures which make up the essential peculiarities of organisms."[14]

Modern biology has also disclosed that life appears to be a prop-erty of the whole that exists within the parts, and the whole is, there-fore, greater than the sum of parts. As Ernst Mayr put it, living sys-tems "almost always have the peculiarity that the characteristics of the whole cannot (not even in theory) be deduced from the most com-plete knowledge of components, taken separately or in other partial combinations. This appearance of new characteristics in wholes has been designated emergence."[15]

The concept of emergence essentially recognizes that an assem-blage of parts in successive levels of organization in nature can result in wholes that display properties that cannot be explained in terms of the collection of parts. As P.B. and J.S. Medawar put it, "Each higher-level subject contains ideas and conceptions peculiar to itself. These are the 'emergent' properties."[16] Because reductionism requires that we explain properties of a whole organism in terms of the behavior of parts at a lower level, it obliges us to view emergent properties as ir-rational and without cause. If, however, we assume that the whole exists within the parts, emergent properties at a higher level can be viewed as properties of a new whole that exists in more complex rela-tion to biological life.

From this perspective, organisms are not mixtures or compounds of inorganic parts but new wholes with emergent properties that are embedded in or intimately related to more complex wholes with their own emergent properties. At the most basic level of organization, quanta interact with other quanta in and between fields, and funda-mental particles interact with other fundamental particles to produce the roughly hundred naturally occurring elements that display emer-gent properties that do not exist in the particles themselves. The parts represented by the elements combine to form new wholes in compounds and minerals that display emergent properties not pres-ent in the elements themselves. For example, the properties in salt, or sodium chloride, are novel and emergent, and do not exist in so-dium or chloride per se.

The parts associated with compounds and minerals combined to form a new whole in the ancestor of DNA that displays emergent properties associated with life. During the first two billion years of evolution, it was the exchange of parts of DNA between prokaryotes and mutations within parts that resulted in new wholes that dis-played new emergent properties. Combination through synergism of these parts resulted in new wholes in eukaryotes that display emer-gent properties not present in prokaryotes.

Meiotic sex, or the typical sex of cells with nuclei, resulted in an exchange of parts of DNA that eventually resulted in new wholes with emergent properties in speciation. Recombinations and extensions of the parts resident in all parts (DNA) resulted in emergent properties in whole organisms that do not exist within the parts or in the series of nucleotides in DNA. Through a complex network of feedback loops the interaction of all organisms as parts resulted in a whole, biological life, which exists within the parts and displays emergent regulatory properties not present in the parts. In the absence of any scientific description of the actual dynamics of the relationships between these levels of organization, however, this understanding of emergent order is not scientific. It is rather a paradigm or heuristic that might occasion more insights that could lead to an improved scientific description.

Emergence in the Whole of the Biota

The fossil record indicates that the temperature of Earth's surface and the composition of the air appear to have been continuously regulated by the whole of life or the entire biota. Although the complex network of feedback loops that maintains conditions suitable for the continuance of life is not well understood, much evidence suggests that the entire biota is responsible. For example, the stabilization of atmospheric oxygen at about twenty-one percent was achieved by the whole biota millions of years ago and has been maintained ever since.

If the oxygen concentration were only a few percent higher, the volatile gas would cause living organisms to spontaneously combust. If it had fallen a few percent lower, aerobic organisms would have died from asphyxiation. This whole also appears to have prevented nitrogen and oxygen from degenerating into substances that would have poisoned the entire system—nitrates and nitrogen oxides. As Margulis and Sagan explain, "If there were no constant, worldwide production of new oxygen by photosynthetic organisms, if there were no release of gaseous nitrogen by nitrate- and ammonia-breathing bacteria, an inert or poisonous atmosphere would rapidly develop."[17]

If we fail to factor in the self-regulating emergent properties of the whole of the biota, the mixture and relative abundance of gases in the atmosphere makes no sense on the basis of chemistry. Oxygen gas forms about twenty-one percent of the atmosphere, and the relative disequilibria of other gases, such as methane, ammonia, methyl chlorine, and methyl iodine, is enormous. If the whole of the biota did not display emergent properties that regulated these parts, chemical analysis suggests that all of these gases, which readily react to oxygen, should be so minute in quantity as to be undetectable. However, nitrogen is ten billion times more abundant, carbon dioxide ten times more abundant, and nitrous oxide ten trillion times more abundant than they should be if these parts had interacted without mediation from the whole.

Physics also indicates that the total luminosity of the sun, or the total quantity of energy released as sunlight, has increased during the last four billion years by as much as fifty percent. According to the fossil record, however, the temperature of Earth has remained fairly stable, about 22 degrees centigrade, despite the fact that temperatures resulting from the less luminous early sun should have been at the freezing level. Because the level of carbon dioxide is mediated by cells, one of the emergent properties of the whole of the biota that maintained Earth's temperature was probably regulation of atmospheric levels of this gas.

Competition Versus Cooperation Within Species

Because Darwin assumed that individual organisms, like classical atoms, are atomized, and that the dynamics of evolution, like the universal force of gravity, acted between or outside organisms, there was no logical basis for conceiving of dynamics that operate within organisms (parts) to coordinate the survival of species (wholes). This forced Darwin to conclude that competition for survival between organisms was the rule of nature and that this competition would be more severe between members of the same species. As Darwin put it, "The struggle will almost invariably be most severe between the individuals of the same species, for they frequent the same districts, require the same food, and are exposed to the same dangers."[18]

In the absence of a struggle for existence between species, Darwin assumed that the rate of increase of numbers of single species would be exponential. "Every single organic being," wrote Darwin, "may be said to be striving to the utmost increase in numbers."[19] If this "utmost increase" is not checked with competition for survival from other species, the consequences, in Darwin's view, are easily imagined: "There is no exception to the rule that every organic being naturally increases at so high a rate, that, if not destroyed, Earth would soon be covered by the progeny of a single pair."[20]

Using the example of elephants, Darwin attempted to estimate the minimal rate of increase in the absence of competition with other species. He assumed that a pair of elephants begins breeding at thirty years old and continues breeding for ninety years, and that six young elephants are born during this period. If each offspring survives for one hundred years and continues to breed at the same rate, Darwin calculated that nineteen million elephants descended from the first pair would be alive after 740 to 750 years.[21] He concluded that this natural tendency for species to increase in number without limit is checked by four "external" causes: predation, starvation, severity of climate, and disease.[22]

Large numbers of field studies by ecologists suggest, however, that one of the primary mechanisms that control the growth in number of species is a whole that exists within parts and not forces that act outside the parts. Take the example of Darwin's elephants. In a study of more than three thousand elephants in Kenya and Tanzania from 1966 to 1968, biologist Richard Laws found that "the age of sexual maturity in elephants was very plastic and was deferred in unfavorable situations." Depending on those situations, individual elephants reached "sexual maturity at from 8 to 30 years."[23] Laws also found that females do not continue bearing until ninety years old, as Darwin supposed, but cease to become pregnant at around fifty-five years of age. The primary mechanism that regulates the population of elephants is the "internal" adjustment of the onset of maturity in females, which lowers the birth rate when overcrowding occurs, not the "external" mechanisms of predation and starvation.

Numerous other studies have shown that internal adjustments in the onset of maturity in females regulates population growth in large numbers of species. Linkage between age of first production of offspring and population density has been found in the white-tailed deer, elk, bison, moose, bighorn sheep, ibex, wildebeest, Himalayan tahr, hippopotamus, lion, grizzly bear, harp seal, southern elephant seal, spotted porpoise, stripped dolphin, blue whale, and sperm whale.[24] This linkage also exists in small mammals.[25]

Many animal species also internally regulate populations by varying their litter and clutch size in response to the amount of food available. According to the biologist Charles Elton, "The short-eared owl (*Asio flammeus*) may have twice as many young in a brood and twice as many broods as usual, during a vole plague, when its food is extremely plentiful."[26] Similarly, nutcrackers, which normally lay only three eggs, increase the clutch to four when there are plentiful hazelnuts, the arctic fox produces large litters when lemmings are abundant, and lions bear fewer or more cubs according to the available food supply.[27]

All of this explains why Darwin's view that only external hostile forces regulate the numbers of atomized organisms has lost some currency among biologists. As biologist V.C. Wynne-Edwards noted, "Setting all preconceptions aside, however, and returning to a detached assessment of the facts revealed by modern observation and experiment, it becomes almost immediately apparent that a very large part of the regulation of numbers depends not on Darwin's hostile forces but on the initiative taken by the animals themselves; that is to say, to an important extent it is an intrinsic phenomenon."[28]

Competition Versus Cooperation between Species

Some evidence also suggests that competition for survival between species (parts) is regulated by a whole (an ecology or ecosystem) that is resident within the parts in terms of the evolved behavior of organisms in an ecology or ecosystem. Even very similar organisms in the same habitat display internal adaptive behaviors that serve to sustain the whole when food and other resources are in short supply. One such adaptive behavior involves the division of the habitat into ecological niches, where the presence of one species does not harm the existence of another similar species. For example, the zebra, wildebeest, and gazelle are common prey to five carnivores: lion, leopard, cheetah, hyena, and wild dog. These predators coexist, however, because they developed five different ways of living off the three prey species that do not directly compete with one another. As ethologist James Gould explained:

> Carnivores avoid competing by hunting primarily in different places at different times, and by using different techniques to capture different segments of the prey population. Cheetahs are unique in their high-speed chase strategy, but as a consequence must specialize on small gazelle. Only the leopard uses an ambush strategy, which seems to play no favorites in the prey it chooses. Hyenas and wild dogs are similar, but hunt at different times. And the lion exploits the brute-force niche, depending alternately on short, powerful rushes and strong-arm robbery.[29]

Herbivores also display evolved behavior that minimizes competition for scarce resources in the interests of sustaining other life forms in the environment. Paul Colvinvaux has studied such behavior on the African Savanna:

> Zebras take the long dry stems of grasses for which their horsy incisor teeth are nicely suited. Wildebeest take the side-shoot grasses, gathering with their tongues in the bovine way and tearing off the food against their single set of incisors. Thompson's gazelles graze where others have been before, picking out ground-hugging plants and other tidbits that the feeding methods of the others have overlooked and left in view. Although these and other big game animals wander over the same patches of country,

they clearly avoid competition by specializing in the kinds
of food energy they take."[30]

Similarly, three species of yellow weaver birds in Central Africa
live on the same shore of a lake without struggle because one species
eats only hard black seeds, another soft green seeds, and the third
only insects.[31] In North America, twenty different insects feed on the
same white pine—five eat only foliage, three live off birds, three on
twigs, two on wood, two on roots, one on bark, and four on cambium.[32]
A newly hatched garter snake pursues worm scent over cricket scent
and a newly hatched green snake in the same environment displays
the opposite preference, but both species of snake could eat the same
prey.[33]

The order that exists within the parts (species) and that appears
to manifest as emergent regulatory properties in wholes (ecosystems)
seems particularly obvious in plants. Each plant in the same environment typically specializes in distinct niches: Some thrive in sandy
soils, others in alkaline, and some require no soil, such as lichens.
Some grow early in the season and others late, and some remain
small and others become huge. In studies of two species of clover in
the same field, one grew faster and reached a peak of leaf density
sooner and the other grew longer petioles and higher leaves that allowed it to overtop the faster growing species and avoid being shaded
out.[34]

While emergent cooperative behaviors within parts (organisms)
that maintain conditions of survival in the whole (environment or
ecosystem) appear to be everywhere in nature, the conditions of observation are such that we distort results when we view any of these
systems as isolated. All parts (organisms) exist finally in an embedded relation to the whole (biota) where the whole seems to operate in
some sense within the parts. As Lynn Margulis explained:

> All organisms are dependent on others for the completion
> of their life cycles. Never, even in spaces as small as a cu-
> bic meter, is a living community of organisms restricted to
> members of a single species. Diversity, both morphological
> and metabolic, is the rule. Most organisms depend directly
> on others for nutrients and gases. Only photo- and chemo-
> autotrophic bacteria produce all their organic require-
> ments from inorganic constituents; even they require food,
> gases such as oxygen, carbon dioxide, and ammonia, which
> although inorganic, are end products of the metabolism of
> other organisms. Heterotrophic organisms require organic
> compounds as food; except in rare cases of cannibalism,
> this food comprises organisms of other species or their re-
> mains.[35]

When we consider that emergent properties of the whole (biota) appear to have consistently maintained conditions for life by regulating large-scale processes like global temperature and relative abundance of gases, the idea that this whole exists within all parts (organisms) becomes rather imposing. Traditional metaphors for the cooperative aspects of life, such as "chain of being" and "web of existence," suggest that the self-regulating properties of the whole are external to or between parts. The more recent metaphor of life as a single organism or cell is a distortion in that it implies that there is no separate existence of parts. Perhaps the more appropriate view is that the relationship between parts (organisms) and whole (life) is complementary.

When we observe behaviors of parts (organisms), the very act of observation necessarily separates the parts from the whole (life). If we attempt to explain all the embedded relations between parts (organisms) and whole (life), the parts become progressively embedded in relation to larger wholes until we reach the event horizon of the whole that is all of biological reality. Although we can, for example, observe the behavior of the whole that is a living cell and the embedded relations of that whole as part with larger wholes (hemoglobin, tissues, organs), these wholes as parts exist in still larger wholes (bodies) embedded in still larger wholes (environments or ecosystems) and so on. If the emergent behavior of wholes could be explained in terms of the assemblage of isolated parts, it would be theoretically possible to observe and represent the whole as the ultimate assemblage of all constituent parts. But it seems clear that the emergent behavior of wholes in organic life cannot be explained in terms of the assemblage of parts, or relations between parts, and is associated with the existence of wholes within parts.

If we analyze the conditions for observation required for unambiguous description, the observation of any collection of parts necessarily precludes observation of any whole where emergent properties cannot be explained in the absence of embedded relations with larger wholes. However, the attempt to observe those relations invokes the existence of progressively more relations to unobserved parts with emergent behaviors that can only be explained in terms of the existence of wholes within those parts. The ultimate extension of this analysis eventually forces us to confront the whole of life that appears to exist within the parts, and yet the existence of this whole cannot be disclosed as any collection of parts no matter how many parts are observed and configured.

Obviously, what we are saying here about the relation between part and whole in biological life is analogous to what we have said about the part-whole complementarity disclosed by nonlocality. In both cases, the whole exists within parts, the whole cannot be disclosed through observation of parts, and the decision to observe parts necessarily separates parts from whole. Because the behavior of parts exists in embedded relation to the whole, both complementary aspects of the total reality must be kept in mind in all acts of observation and in the analysis of relations between parts. However, in the absence of any understanding of mechanisms linking quantum mechanical processes and progressive emergent behavior in biological life, the only valid conclusion is that the logic of complementarity can serve as a heuristic for understanding fundamental part-whole complementarities in both physical and biological realities.

It is important to realize, however, that even if we do discover a linkage between quantum mechanical processes and emergent behavior in biological life, this will not result in a one-to-one correspondence between physical theory and physical reality in either physics or biology. Suppose, for example, that we construct an enormously elaborate computer model of all the variables that might account for the symbiosis and cooperation abundantly evident in Earth's ecosystem. Our impulse would be to isolate the system called "life" by modeling its dynamics within the larger life system that is the ecosystem. Would this impossibly elaborate program allow us to fully explain the mechanisms of symbiosis and cooperation as well as competition between species?

It could not. Most obviously, the ecosystem, like any system, cannot be isolated from the rest of the cosmos in accordance with modern physical theory. Suppose, however, we seek to obviate that problem with the argument that because we are dealing with macro-level processes, the speed of light and the quantum of action need not concern us in arriving at practical or workable results. This argument will not save the conditions for our isolated experimental situation for a simple reason. The indeterminacy of quantum mechanical events inherent in every activity within the ecosystem would become a macro-level problem when dealing with a system on this scale.

In the next two chapters, we will explore how this new understanding of the relationship between parts and wholes in physical reality can serve as a heuristic that could lead to future advances in the study of cosmology. We will begin with a brief history of cosmology that will serve as background for understanding problems associated with the current model for describing the origins and history of the cosmos—the big bang model with inflation.

6
Ancient Whispers: The Expanding Universe

For years, the community has worked to understand the deepest sounding of the universe, a universal background far beyond the stars and even the galaxies which fills the sky: the most energetic and the most ancient signal we have ever seen.

Philip Morrison

The story that we now tell about the origins and history of the cosmos is perhaps the most dramatic demonstration of the ability of mathematical physics to coordinate experience with physical reality. The fact that this physics, which originated a mere instant ago in cosmic time, allows us to consider how the entire universe began and the manner in which it evolved over billions of years staggers the imagination. The current model for describing this universe, the big bang model with inflation, has enjoyed great success, but there are many indications that it may be badly flawed. To better understand how the original elegance and simplicity of this model has been compromised, we will begin with a brief history of cosmology.

In the seventeenth and eighteenth centuries, Newtonian dynamics and the law of universal gravitation allowed the motions of the planets and stars to be understood for the first time by using well-founded physical theory rather than philosophically determined mathematical constructs such as Ptolemy's epicycles. Although observations eventually indicated that the universe was much larger than anyone could have imagined, classical physics could not say anything about its origins. Based largely on aesthetic and metaphysical concerns, Newton assumed that the universe was infinitely extended in space and static, or eternally the same. Einstein was also drawn to the aesthetic and/or metaphysical appeal of a static universe. In his first general relativistic model, the universe was assumed to be unbounded, finite, and spherical, but it was also assumed to be static or eternally the same. Even after Einstein published his general theory of relativity in 1915, he did not anticipate that it would become the basis for positing the existence of a nonstatic, evolving model for the universe. With the theoretical framework of general relativity already in place, the Belgian cosmologist Abbé Lemaitre and the Russian mathematician Alexander Friedmann postulated a model of the universe that was dynamic and evolving.

The first convincing evidence that the universe was not static, and that the Friedmann-Lemaitre model was more than a theoretical curiosity, came from observational astronomy. Using new instruments probing deep space, such as the 100-inch telescope at Mt. Wilson and the 200-inch telescope at Mt. Palomar, astronomers broadened the horizon of the observable universe to billions of light years. In the 1920s, Edwin Hubble used the Mt. Palomar telescope to gather evidence that strongly indicated that the universe was expanding and evolving.

The early relativistic models of Friedmann and Lemaitre were, however, premised on a set of assumptions, reflected in the so-called "cosmological principle," that are essentially Newtonian. This principle states that the universe must be isotropic, meaning the same in all directions, and homogeneous, meaning that the density of the universe is the same on average at all points in space. Assuming that the total amount of energy-matter is constant, the model suggests that the average density of matter in the universe over time becomes lower as the universe expands and the distribution of matter becomes more dilute.

The appeal of a static, eternal universe is also apparent in the steady-state theory advanced by Herman Bondi, Thomas Gold, and Fred Hoyle in the 1940s. Although the model features an expanding universe, it assumes that the universe appears the same to all observers at all times because it obeys the so-called "perfect cosmological principle." While the model attempted to accommodate the observational evidence supporting an expanding universe, it did so within the framework of an eternal universe, which was apparently more philosophically palatable to its founders. The price paid to accomplish this feat was the abandonment of a foundational principle in physics—the conservation of mass and energy. Matter in this model is created continuously *ex nihilo,* or from nothing, and space is filled with this new matter as the universe expands. Hence the average density of the universe always remains the same and, therefore, the universe always looks the same. The steady-state model, for obvious reasons, is also termed the "continuous creation model."

What these ad hoc assumptions allowed one to do was to avoid, or obviate, a central question: how did the universe come into existence?

An evolving universe without continuous creation of matter requires that the average density of the universe must get lower over time. If we run the clock backward and reverse this process, it would be clear that the universe must have been much denser and hotter much earlier in time. If we imagine that the expansion is reversed long enough, as Lemaitre, Friedmann, and others did in the 1920s, this suggests that the universe originated at a single point called the "primordial singularity," or as Lemaitre termed it, from a "primordial *atom*."

The Hot Big Bang Theory

In the early 1950s, the cosmologist George Gamow extended Lemaitre's and Friedmann's original ideas by appealing to quantum physics. At this time, advances in atomic and nuclear physics had began to provide some insights into the early life of the universe, where energies and temperatures associated with the domain of violent nuclear collisions and high energy physics would be apparent. If the label for this model—the "big bang"—seems less than dignified in describing how the grandeur of the cosmos came into being, there is a reason why this may be the case.

The term "big bang" comes from Fred Hoyle, who advocated the opposing steady-state model, and it was not intended to dignify Gamow's theory. For several years, however, it was not clear which of the two views would eventually triumph. While the steady-state theory was appealing in its elegance and simplicity, the universe in this model was without origins. Meanwhile, observational astronomy, armed with the new branch of millimeter and radio astronomy and sophisticated deep space optical spectroscopy, provided evidence in favor of the big bang model. Among this evidence, there were two discoveries that strongly indicated that the big bang model is correct. The first was the 3 degrees or, more accurately, the 2.73 degrees Kelvin, black body radiation, which was immediately understood as the most ancient whisper of the origin of the cosmos; the second was a new class of distant astronomical objects known as "quasars."

The microwave black body radiation, accidentally discovered by Bell Telephone physicists Arno Penzias and Robert Wilson in 1965, had been predicted by Gamow for some time. Gamow assumed that a background of radiation that fills all space would represent the relic or whisper of primordial radiation from the initial big bang and that this radiation would fill a greater volume of space as the universe expands and cools (see Fig. 20). Because the total energy content of the universe is conserved, the energy density, or amount of energy of radiation per unit volume, must, he said, decrease. This means that what used to be a hot background of radiation filing all space, composed of X-rays and gamma rays, would be redshifted over time to the microwave region of the spectrum and would eventually acquire a much lower temperature.

The shift to the red is used to calculate the rate at which a luminous body is moving away from us in space. When spectral lines in the light from an object like a star are shifted toward the red, or long-wavelength end of the spectrum, we can calculate the speed at which the object is moving away from Earth. This is also known as the Doppler effect, after the Austrian physicist who first described it in 1842; it can be observed in sound waves when an ambulance with its siren on approaches and passes us on the highway. The sound waves of the siren are shorter and louder as the ambulance approaches and become longer and less loud as the ambulance moves away.

Because the big bang theory could explain the existence of primordial radiation (and may have even predicted its existence), this theory appeared victorious in competition with the steady-state theory. In steady-state theory, which does not allow origins, the microwave background must presumably arise from many distant sources that somehow add up to produce an unresolved background of radiation that mimics black body radiation. No one, however, was able to discover suitable candidates for these alleged sources. Although the microwave radiation is a natural outcome of the big bang model, competing theories were obliged to explain it with assumptions that were often ad hoc.

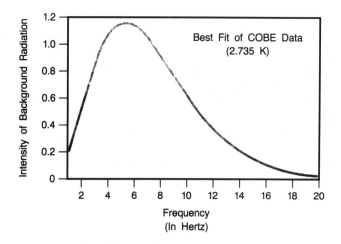

FIGURE 20. Data from COBE showing that the cosmic background radiation fits a black body curve at 2.735 K remarkably well.

The second major blow to steady-state theory was the discovery in the 1960s of very distant objects called quasars (see Fig. 21). To appreciate how distant these objects are, we must first remind ourselves that the maximum distance from Earth where receding galaxies can be observed is a function of the speed of light. If the speed of recession of celestial objects becomes equal to the speed of light, these objects cannot be observed from Earth. This limit on observation is known as the "horizon of the universe." The distance to other galaxies within this horizon is determined by the extent to which light from a galaxy has been redshifted. It was this phenomenon that became the basis for the Hubble law—the more distant a galaxy, the more its spectrum is shifted to lower frequencies or longer wavelengths.

What was unusual about the spectral lines from the quasars, or "quasi-stellar objects" (QSOs), is that they indicate that the light has been redshifted to very low frequencies compared with light from other celestial objects. The light of quasars is redshifted to such low frequencies, compared to that of other astronomical objects, that we must presume that these objects are near the very edges of the observable universe. These objects are typically at eighty to ninety-five percent of the distance to the edge of the horizon, and some are even greater than that. Some quasars are so distant from us that they seem to be receding, or moving away from us, at speeds exceeding ninety-five percent of the speed of light.

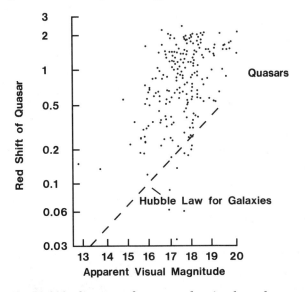

FIGURE 21. The Hubble diagram of quasars showing how the speed of recession of a quasar measured by its redshift correlates with the distance of the quasar. Quasars do not obey the law followed by galaxies. Data collected by astronomer Dr. E.M. Burbidge.

Eventually, it was understood that quasars are very distant galaxies and that the light from these galaxies has been traveling for billions of light years. Because light traveling such distances normally becomes so faint that the source cannot be observed, it was immediately realized that there was something unusual about light from quasars. The brilliant, star-like, nucleus of the galaxies from which the light of the quasars originated must have been far more brilliant than the nucleus of nearby galaxies; this suggests that the universe does not, in fact, look the same at all times to all observers.

If galaxies were much brighter in the past than they are today, then the perfect cosmological principle, which is foundational to the steady-state theory, does not hold. The apparent disagreement of steady-state theory with observations finally prompted most of its supporters to abandon it soon after the discovery of quasars. What was not anticipated, however, is that improved modern cosmological observations, which are often conducted on platforms in space, would pose challenges to the big bang theory. The irony is that the observations that helped undermine the steady-state theory and establish the pre-eminence of the big bang theory eventually produced new constraints and complications for big bang theory itself.

Advances in quantum field theory have afforded us a unique opportunity to describe the conditions of the universe near the big bang. This was not possible a few decades ago because the extremely high temperatures and densities prevalent in the early universe required answers from particle physics that were totally unknown. In our new situation, particle physicists and cosmologists can test predictions about conditions in the early universe using high-speed particle accelerators. All the evidence now suggests that the very early universe, or the universe as it existed in the first fractions of a second, was in a quantum state, and the consensus is that a quantum theory of gravity is needed to move toward a more complete and self-consistent cosmology. Progress in this area over the last few years has been quite stunning due largely to the success of the quantum field theory of particle interactions.

Problems with Big Bang Theory

As observations have moved cosmology from the realm of philosophical speculation to the realm of testable physical theories, several problems have emerged: The three main problems faced by the hot big bang model are known as the flatness problem, the horizon problem, and the isotropy problem. The flatness problem has to do with the fact that the original big bang theory cannot account for the observed density of matter in the universe. What is curious about this observed value is that it is quite close to what would be required to "close" the universe.

Quantitatively, this is expressed as $\Omega = 1$, where Ω is the density ratio $\Omega = \rho / \rho_c$ and the critical density can be expressed as[1]

$$\rho_c = 2 \times 10^{-29} (H_o / 100 \text{ km/sec/Mpc})^2$$

where H_o is the Hubble constant that measures the rate of expansion of the universe. If $H_o = 100$ km / sec / Mpc, this means that as the distance increases 1 megaparsec (Mpc), or three million light years, the rate of expansion of the universe increases by one hundred kilometers per second. For $H_o = 100$ km / sec / Mpc, the horizon of the universe lies ten billion light years away from Earth. The precise value of H_o remains the most fundamental challenge of observational cosmology. Although most observers believe that it lies in the range 50–100 km / sec / Mpc, recent observations using the Hubble Space Telescope indicate that it might be in the middle of this range. This suggests that the horizon is twenty to ten billion light years away, and that the universe is approximately twenty to ten billion years old.

If the value of the density ratio turned out to be precisely equal to unity, the geometry of the universe would be exactly flat and the universe would expand until it finally comes to rest in the very distant future. If the density turns out to be less than the critical value, the geometry of the universe would be similar to the surface of a saddle. This would mean that the universe is open, and, like the flat universe case, its expansion would never cease. Suppose, however, that the density turns out to be greater than the critical density. The geometry in that event would resemble the surface of a sphere, and the universe would be closed. In this case, the universe would expand to a maximum size and then start collapsing back onto itself in the so-called "big crunch" (see Fig. 22). It is important to realize here, however, that the density of the universe is extremely close to the closure or critical density.

The mean density of the universe is computed by estimating the masses of distant galaxies, based on their observed brightness, and comparing those values with what is known about nearby galaxies. When all the luminous matter in a given volume of space is added up, the density of luminous matter is determined. The obvious problem is that current observations cannot unequivocally determine the type of universe we live in.

Although most observers favor values close to 0.1 or less, values of Ω for dark matter and luminous matter allowed by the Hubble diagram of distant galaxies in the range 0.1 to 2 are possible.

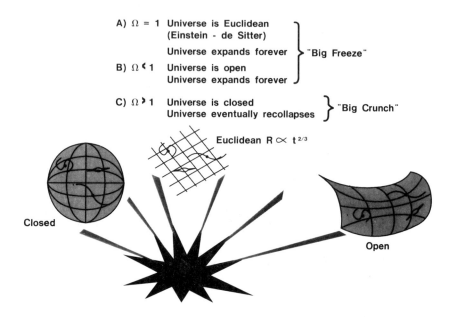

A) $\Omega = 1$ Universe is Euclidean
(Einstein - de Sitter)

Universe expands forever

B) $\Omega < 1$ Universe is open
Universe expands forever

$\Bigg\}$ "Big Freeze"

C) $\Omega > 1$ Universe is closed
Universe eventually recollapses

$\Bigg\}$ "Big Crunch"

Euclidean $R \propto t^{2/3}$

Closed

Open

FIGURE 22. The geometries of the three types of big bang cosmological models, open (< 1), closed (>1), and flat (=1).

If the only type of matter in the universe is luminous matter, like that found in stars and nebulae, the results would indicate that the universe is open. However, the amount of matter in all stars, intergalactic gas, and baryons required to produce the observable elements is much too small. It does not appear to be much greater than $\Omega \sim 0.04$. Astronomers suspect, however, that there is much more dark matter in the universe than all the luminous matter making up the galaxies. It is conceivable that more than ninety percent of the matter in the universe that the big bang flat universe cosmology requires is in the form of dark, invisible matter. Although the nature of this matter is controversial, most theorists think it is composed of cold, dark matter in the form of particles that have not yet been detected in the laboratory. However, because there have been numerous failed attempts to detect the existence of this matter, we must continue to view the existence of this matter as a theoretical speculation that may not be in accord with scientific facts. Even though we have recently discovered that neutrinos have mass, this has not improved the situation. The point is that attempts to discover the existence of the large amount of matter have failed, and estimates of the contribution of dark matter to Ω generally fall in the range ~ 0.2–0.5.

Although current observations provide only an approximate range for the mean density of the universe, what is perplexing is that the value is so close to the critical density required for a flat geometry. Even if it turns out that the universe is close to flat today, it must have been incredibly flat in its early stages to one part in 10^{50}. The reason for this conclusion is straightforward. If the universe, which has expanded by 60 orders of magnitude since the big bang, is approximately flat now, this would translate into a condition of rather exact flatness in its early stages.

The horizon problem relates to the uniformity of the black body radiation that emerged about 1,000,000 years after origins. The apparent homogeneity of this radiation was recently observed by the Cosmic Background Explorer, or COBE.[2] Because the radiation is so uniform that it has the same temperature to within one part in one hundred thousand in every direction on the sky, one must assume that opposite parts of the sky were in causal contact when it was first emitted. This appears, however, impossible because opposite parts of the sky in the expanding universe at the point at which the radiation first appeared were separated, or out of causal contact, by $\sim 10^7$, or ten followed by seven zeros, light years. Because no signal, according to the theory of relativity, can travel faster than light, we must assume that the causal contact between regions in space required to explain the uniformity of the background radiation under normal conditions did not exist.

The last problem that the original big bang model cannot explain is the isotropy problem, or the fact that the universe appears fairly smooth or homogeneous in all directions. Given the large number of possibilities in the initial conditions allowed by the model, one would not expect the universe to behave in such a simplified manner. Assuming that the universe started from a hot big bang, it appears that that the total chaos that prevailed in the early universe would not have died away.[3] In fact, quantitative calculations show that slight anisotropies, or variations from the mean uniformity of matter, would have been greatly amplified.

The view of the universe as isotropic was considerably reinforced, as we saw earlier, by the discovery that the background microwave radiation appears remarkably uniform no matter where one examines it in space. Because the universe is presumed to be expanding the same way in all directions, it was reasonable to conclude that matter, like the background radiation, should also be isotropic or remarkably uniform in all directions in deep space. The expectation was that at some point in deep space we would begin to observe uniformity in the density of matter. The problem is that this has not been observed where, according to the big bang model, it should be apparent. There is, for example, no such uniformity of matter in the superclusters that are hundreds of millions of light years or more in extent.

It seems clear at this point that the universe is not homogenous at all scales and in all directions. As astronomers look further into space, some incredibly large structures have been observed (see Fig. 23). One such structure, known as the Pisces-Cetus supercluster, extends to about ten percent of the observable radius of the universe.[4] Galaxies now appear to cluster themselves in increasing hierarchies of clusters, called "superclusters." Beyond this level, superclusters may form even larger structures, supersuperclusters, that approach a sizable fraction of the radius of the observable universe. The observable largest superclusters often assume the shape of filaments that appear to lie on the surface of very large bubbles of empty space termed the "voids." Galaxies have also been observed that do not move in the isotropic fashion required by the uniform expansion of the universe.[5] What is perhaps more surprising, based on the totality of what is known in contemporary cosmology, is that the universe is not more anisotropic, or less homogeneous, than the observations suggest.

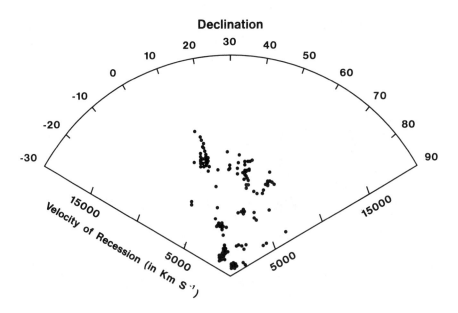

FIGURE 23. The Hercules Supercluster is a large structure of many galaxies that seemingly violates the condition of homogeneity of matter in the universe.

Inflation as the Solution to the Big Bang Problems

The current big bang theory of the universe attempts to eliminate the flatness problem, the horizon problem, and the isotropy problem with the so-called "inflationary hypothesis." According to this hypothesis, which was first proposed by Demos Kazanas, Alan Guth,[6] and others, the universe emerges from the vacuum state, or a completely unvisualizable "point nothing" that can generate enormous energy. In the first phase, or at the initial singularity, which physics cannot describe, the space-time description breaks down entirely. Because Planck first calculated the smallest dimensions where space-time would break down, this era is named after him.

This phase is allegedly followed by the "inflationary era" at about 10^{-35} sec.[7] During the proposed inflationary era, the universe presumably underwent an extremely rapid expansion and doubled in size every 10^{-35} of a second by a staggering factor of 10^{50} or more. The process through which water freezes provides a crude analogy for understanding what allegedly occurred during the inflationary era. Like the freezing of water, inflation involves a phase transition out of a false vacuum that contains tremendous energies.

Prior to inflation, the universe is viewed as being in a phase of symmetry with respect to the so-called "Higgs fields," and the strong, weak, and electromagnetic interactions were unified. The Higgs fields are members of a special set of quantum fields postulated in grand unified theories (GUT), and they serve to account for the spontaneous symmetry breaking that would lead to their emergence. The state of unbroken symmetry when the Higgs fields were zero, or prior to the onset of inflation, is known as the false vacuum that contained within itself tremendous energies. The numbers here are quite impressive— every cubic centimeter of volume in this false vacuum would have to contain 10^{95} ergs of energy, or about 10^{19} more energy than the mass converted into energy of the entire observable universe.

As this early universe expands and cools to a temperature of about 10^{27} K, Guth and others proposed that a phase transition begins from the false vacuum, where all the Higgs fields were zero, to a less energetic phase termed the "true" vacuum. In this true vacuum state, the Higgs fields acquire nonzero values, and the GUT symmetry breaks down. At this point the universe allegedly enters an energetically stable phase, meaning that the amount of energy available to it for the remainder of its existence was fixed. The theory also indicates that the breaking of the GUT symmetry in the expanding universe allowed the strong force to separate from the electroweak force.

In normal gases, the pressure is positive, and an aggregate of hot gases, like that of a star, would tend to expand as gravity tends to pull it together. The false vacuum during the inflationary era is understood, however, as having a very large negative pressure. Although there was no ordinary hot matter and no ordinary pressure, we are obliged here to conceive of a tremendous negative pressure coming out of the nothingness of the false vacuum. Under these conditions, general relativity predicts that gravity, rather than pulling the early universe together, would have the opposite effect of pushing it away. Hence the universe inflates.

The inflationary model requires that we view the universe as expanding over a period of 10^{-30} sec from a size of less than 10^{-50} of a centimeter to about one centimeter. This suggests that the universe expanded by more than fifty orders of magnitude in an incredibly small fraction of a second. The subsequent expansion of the universe to its currently observed size of roughly 15 billion light years is trivial when compared to the staggering expansion called for in the inflationary model. As the universe was allegedly moving from the nothingness of the false vacuum to the nothingness of the true vacuum, the model indicates that the tremendous energies locked up in the negative pressure of the false vacuum were released. It is inconceivable, as the inflationary model requires, that all of the observable matter in the universe came out of the nothingness of the false vacuum. The problem, however, is that it also implies that the universe should be much larger than what we observe it to be.

What is most important to realize about the inflationary model is that it was not originally proposed for any compelling theoretical reasons but rather to solve the three observational problems faced by the standard big bang theory. Inflation seeks to solve the flatness problem by assuming that flatness would be a natural consequence of rapidly inflating the curvature of space-time. No matter what curvature one starts with, the expansion of space-time by a factor of 10^{50} or more would create a universe that was remarkably flat. If one, for example, could inflate a balloon by such a factor, the surface of the balloon would closely approximate a flat surface.

The model appears to resolve the horizon problem because it allows us to assume that the regions on opposite sides of the expanding universe were in causal contact after the singularity and before they were blasted apart by inflation. Inflation does not solve the isotropy problem, and there are some recent indications that it will face serious problems in the effort to do so. But even if inflation appears to solve this problem, the big bang model with inflation requires a horizon that is considerably larger than the size of the universe based on current observations. The universe required in this model would, in fact, have to be ten million billion billion billion times larger than we observe it to be.

The remainder of the big bang story is less controversial. Between 10^{-30} and 10^{-6} sec, the universe was filled with a primordial soup of quarks and light particles, like the electrons. About 10^{-6} second after the big bang, the quarks combined to form the heavy particles, like protons and neutrons. In the next phase of ascending complexity, between 1 second and 3 minutes, protons and neutrons underwent nuclear reactions forming nuclei of helium and its isotopes. Although some heavy elements were formed during this period, their amounts were minute. The formation of appreciable large amounts of those elements occurred millions of years later in massive stars that became exploding supernovae. These elements were spewed outward into space by these powerful stellar explosions.

Prior to about 100,000 years after the big bang, the universe was so hot and dense that photons and matter were coupled together. About this time, however, the universe expanded to the point at which photons decoupled and the 2.7 K black body photons escaped at this time. What is important to realize at this point in the story is that these photons provide no opportunity to probe periods prior to 100,000 years, where fundamentally important quantum processes were occurring. This means that the main tool of observational astronomy, light quanta, simply cannot be used to verify what occurred during this crucial early quantum phase of the universe. Other means that we have of probing this period, like estimating the amount of primordial helium, are not as direct. What we have here is a situation in which the limits that observations impose on the theory are intrinsic; they will not, in other words, go away with future improvements in observations.

Recent results obtained with the Hubble Space Telescope indicate that the universe is not deaccelerating as fast as if inflation had actually occurred. Instead, observations of distant supernovae indicate a negatively curved or open universe. The only way to reconcile these observations with a flat universe is to reintroduce the cosmological constant of Einstein, which would require fine tuning of one part in 10^{120}, or much finer tuning than the inflation model allows. Structure formation in an inflationary universe requires the contribution of dark matter, most of it outside the observable galaxies, to Ω to be between 0.2 and 0.5.[8] This implies that the contribution of the cosmological constant to Ω is ~ 0.5 – 0.8. Here, if we want to preserve the whole structure of the inflationary, dark matter-dominated hot big bang, then the least-understood and most speculative concept encountered in the cosmology—the mysterious cosmological constant that Einstein

7

The Emergence of a New Vision: The Unfolding Universe

> In the world of quantum physics, no phenomenon is a phenomenon until it is a recorded phenomenon.

> *John A. Wheeler*

The central thesis in this chapter is that the universe must be viewed as a quantum system not merely in its very early stages following the origin but at all scales and times. If this thesis is correct, we must adopt some new epistemological assumptions in the study of cosmology, or of the origin, history, and large-scale structure of the universe. As we have seen, the resolution of the observation problem in quantum physics required an understanding of the epistemological assumptions in the mind of the observer that are directly related to a proper interpretation of the results of experiments. This also applies in our view to the resolution of critical observation problems in contemporary cosmology examined before. The intent is to illustrate the manner in which classical assumptions have functioned as distorting lenses that conditioned our view of the origin and history of the cosmos in fundamental ways.

All the evidence indicates that the universe is extremely fine-tuned, and it could be that this fine-tuning involves linkages between microscopic and macroscopic quantities. One of the most striking indications of this is the appearance of certain coincidences in quantities that, for all practical purposes, should not be even be related to each other. The mathematical physicist Hermann Weyl was the first scientist to notice that a very larger number, 10 followed by forty zeros, described very disparate quantities. In 1919, Weyl discovered that 10^{40} appeared in the ratio of the electron radius to the gravitational radius of its mass. The suggestion that this number might have special value also influenced Arthur Eddington in his life-long quest to determine precise values for the constants of nature. Although his estimate of the number of particles in the universe was 10^{79}, the fact that the value was within a range of the square of 10^{40} by a factor of a hundred was intriguing.[1]

In 1937, Paul Dirac attempted to express the age of the universe in a measure that was less arbitrary than a celestial year and decided that atomic time units of the proton and electron could serve as the basis for a more scientifically valid clock. After determining that these atomic units were 10^{40}, he demonstrated that the reciprocal for the fine structure for gravity yielded the same number. What seemed utterly remarkable was that this incredibly large number appeared in different branches of physics. The age of the universe is a cosmological problem, the gravitational constant is a property of the cosmos at the micro level, and atomic time units are a micro-level dynamic derived from quantum physics.

The following describes the ratio of the electric forces to gravitational forces:

$$e^2/Gm_e m_p \sim 10^{40}$$

The ratio of the observable size of the universe to the size of an elementary particle is:

$$R/\left(e^2/m_e c^2\right) \sim 10^{40}$$

where the numerator or scale of the universe is changing as it expands.

According to Dirac's large number hypothesis, the fact that these two ratios are equal for all practical purposes is not a mere coincidence, and a number of attempts have been made to demonstrate that this is the case. The possibility that constants, such as the gravitational constant, may be varying was proposed by Dirac[2] and others.[3] Other ratios, such as the ratio of an elementary particle, the electron, to the Planck length also suggest that there may be some deep underlying harmonies involving the fundamental constants that link the microcosm to the macrocosm in fundamental ways.[4] Although theory has not yet accounted for these harmonies in a self-consistent way, this enigma could be resolved by a quantum theory of gravity that unifies all physical forces at the quantum level or by advances in superstring theory. If this is the case, we will have another demonstration that remarkable similarities between physical quantities can lead to new insights into the life of the cosmos despite the fact that the classical explanation for this correspondence is no longer tenable.

The size of an elementary particle to the Planck length:

$$\frac{e^2/m_e c^2}{\left(\eta G/c^3\right)^{1/2}} \sim 10^{20}$$

The Universe as a Quantum System

It is conceivable that all of the observation problems associated with the big bang model can be eliminated by viewing the universe as a quantum system. Although this assumption is more in accord with the totality of what is known about the universe, the proposed solution is not as simple as might first appear. If the universe is a quantum system, it is obvious, first of all, that we can no longer treat it as a closed system separate and discrete from the observer. Hence, observations of even the largest system known to us, the universe, must include the observer and his measuring instruments. Although this view suggests that all acts of observation in cosmology should be theoretically subject to this condition, we will confine ourselves to discussing those in which the rules of observation in quantum physics must clearly be invoked. Because what we are proposing represents a radical revision of normative assumptions in cosmology, we should probably explain why cosmologists, in our view, have tended to resist the idea that the universe is a quantum system.

We think that the primary reasons for this resistance are similar to those discussed earlier in connection with alternatives to Bohr's Copenhagen Interpretation. Although quantum physics has been quite successful in describing conditions in the early universe, cosmologists and particle physicists have, in general, resisted the idea that what was clearly a quantum system in the early stage continued to manifest as such in the later stages. Even though quantum theory is accepted as the only means of arriving at a description of the very early stages in the life of the universe, the description of all subsequent stages in the hot big bang model with inflation is premised on assumptions that come from classical physics. This explains why virtually all cosmologists assume that the early universe was a quantum system and studiously avoid the possible implications of this fact in terms of the role of observer and the continuity of physical processes in all subsequent stages.

One reason physicists have been able to continue to operate in this way is that the general theory of relativity has not yet been quantized. Although this theory has, of course, been highly successful, it is still basically anchored in classical terms. The conceptual framework of general relativity, the space-time continuum, is, as the name implies, a continuum. There is no place in general relativity for discrete quantum jumps and interactions. The manner in which a general relativistic description of the expanding universe can be made compatible with a quantum mechanical description remains an enigma. It is anticipated, however, that a quantum theory of gravity will remove this enigma and if, or more likely when, this occurs, it is hard to imagine how one could resist the conclusion that the entire universe is a quantum system.

Another reason physicists have not been inclined to view the entire universe as a quantum system is more directly related to the legacy of classical physics. In classical physics, as we have seen, parts can theoretically be known with absolute certainty, and the ultimate assemblage of parts constitutes the whole. Because the classical physicist believed in locality and unrestricted causality, the part could be treated as a closed system separate and distinct from the observer. Relying on the laws of conservation of energy, momentum, and angular momentum, classical physics allowed cosmologists to presume that they could know with ultimate certainty the actual state of a collection of parts in any closed system, even if that system is the entire universe.

The assumption of unrestricted causality also led to the presumption that subsequent events in the expanding universe could be known after observations disclosed the initial conditions in the system. It was, therefore, reasonable to assume that if the description of the initial conditions in the universe was sufficiently complete, we could know all subsequent events in the evolution of the universe. Because knowledge of the universe in this sense depends on the existence of unrestricted causality and a one-to-one correspondence between every aspect of the physical theory and physical reality, cosmologists are understandably reluctant to accept the view of the cosmos as a quantum system.

Although cosmologists typically operate on a belief in this correspondence in the search for an experimentally verifiable theory for the origins of the cosmos, they are applying classical assumptions in a situation where they do not by any reasonable criteria make sense. Quantum effects in the very early stages of the life of the cosmos were large, and the most modern of all cosmological theories, the hot big bang model with inflation, speculates that the universe first came into existence as a result of a quantum transition. During the Planck era close to origins at times earlier than 10^{-43} sec, quantum effects were pervasive and inescapable. In the Planck era, even the fundamental construct of space-time is not applicable.

What we are asked to envision here is a universe filled with mini black holes quickly arising from and disappearing into a background of nothingness. The conditions of that era are still present today for sizes approaching the Planck length, or in the unimaginably small dimensions of 10^{-33} cm. Theoretically at least, we could observe effects that were present at the singularity if we could generate sufficient energy to open the door into a past that is still present.

But even if we could conduct this experiment, which would require that we generate energies well beyond any technical capabilities now imaginable, would the information provided about initial conditions allow us to presume understanding of all subsequent events? Obviously not. This is a quantum domain requiring rules of observation, and the quantum of action definitely makes the situation too ambiguous to presume accurate prediction of all future events. In other words, we have an observational problem that exists in principle irrespective of future improvements in observations. If we insist on applying classical presuppositions, our experience in quantum physics clearly suggests that this will only lead to ambiguities that cannot be resolved.

However, dealing with the observation problem in a quantum universe is, as John Bell noted, more than a little problematic. In a laboratory setting, the requirement that we factor into our understanding of results in quantum physics the relationship between the observer and the observed system can be met. But how does one meet that requirement when the observed system is the entire universe? This problem becomes particularly acute in a practical or operational sense in the very early stages of the life of the universe where quantum effects are extremely important.

This problem cannot, however, be ignored. Quantum effects, even though they are quite small and can often be ignored as a matter of convenience, are pervasive throughout the history of the universe. It is, therefore, quantum theory that promises to provide the most complete description of the history of the evolution of the cosmos. One cannot, in theory at least, presume a categorical distinction between acts of observation and the observed system even if the system is in excess of the billion galaxies contained in the observable universe. We, as observers, are clearly included in experimental situations dealing with the largest system imaginable just as we are included in dealing with the smallest possible system in the quantum domain.

The only way we could conceivably know the history of the universe in the classical sense would be to know the sum of all energies, momenta, and other physical quantities of all objects in the universe at any one instant, including at the origin. Quantum indeterminacy obviates that prospect in principle. There is no outside perspective from which to view the physical universe, and any theory based on such a perspective is, in our view, an idealization that cannot be in the final analysis self-consistent.

The Observational Problem in Cosmology

Although most cosmologists view the early universe as a quantum system, they treat its evolution in classical or nonquantum terms.

Most also tend to view the resolution of cosmological observation problems in terms of classical assumptions about the independent existence of macroscopic properties of the observed system. The large problem here is that the principal source of knowledge about phenomena in the early universe is light quanta. If we view the universe as a quantum system, then some ambiguities that have resulted from observations based on light quanta in astronomy might be explainable in terms of quantum indeterminacy. The photographic evidence produced by these observations involves the particle aspect of light quanta, and observations based on spectral analysis involve the wave aspect of these quanta.

If we are observing only a few photons from very distant sources at the edge of the observable universe, the resulting indeterminacy must be imposing some limits on the process of observation.[5] If this is the case, complementarily must be invoked in our efforts to understand the early life of the universe based on observations involving few light quanta due to the ambiguities introduced by the quantum of action. In these situations, the "choice" of whether to record the particle or wave aspect could have appreciable consequences.

What makes this hypothesis reasonable is the fact that observation at the point at which the horizon of observation appears clearly invokes the rules of observation in quantum physics. Because this horizon appears as we attempt to detect fainter sources in the sky, observations near this horizon are necessarily based on fewer photons. Because observation involving small numbers of photons clearly invokes the quantum measurement problem, the choice of which aspect of quantum reality we elect to observe should have appreciable consequences. The only way we can presume to know all the macroscopic variables involved when the observation is based on small numbers of photons is to treat what is clearly a quantum system as a classical system.

The apparent fact that observation in such situations suggests that we must invoke wave-particle duality becomes particularly intriguing when we consider that observational limits appear to be a consequence of adopting "single" and "specific" theoretical models of the universe. Why is it, for example, that the big bang model leads to ambiguities that cannot apparently be resolved within the context of that model? Or why is it that any single model of the universe seems to inevitably result in ambiguities that seem impossible to resolve through observation? The usual answer to these questions is that improved theory and observations will resolve the ambiguities. However, there are now some good reasons to believe that the observational limits, or what we term "horizons of knowledge," are inherent in the laboratory we call the universe and that improved observational techniques simply cannot remove the ambiguities.

The radical suggestion here, which will doubtless disturb many cosmologists, is that these horizons of knowledge testify to the fact that the observed system is a quantum system. If this is the case, then it is conceivable that single and specific models create ambiguities that cannot be resolved within the context of those models because we have failed to consider the prospect that logically disparate models are complementary. The two models that should perhaps be viewed in these terms are the open and closed models of the universe. This thesis is explored in detail by Kafatos.[6]

Although there is no general agreement about the ultimate fate of the universe, virtually all cosmologists and quantum physicists agree that all quanta were entangled in the early universe. If that is the case, we must also conclude that this quantum entanglement remains a frozen-in property of the macrocosm. Quantum entanglement in the experiments testing Bell's theorem reveals an underlying wholeness that remains a property of the entire system even at macroscopic distances. This forces us to conclude that the underlying wholeness associated with quantum entanglement in the early universe remained a property of the universe at all times and all scales.

Nonlocality and Cosmology

If we are correct in assuming that the entire universe is a quantum system and must be regarded as such in cosmology, we are required to draw some dramatic conclusions in epistemological terms. Bell's theorem and the experiments testing that theorem have forced us to embrace a profound new synthesis between the observer and the observed system by confirming the real or actual existence of nonlocality as a new fact of nature. Another experiment that carries large implications in the effort to understand the new epistemological situation in cosmology is the delayed-choice experiment of Wheeler discussed earlier.

Recall that Wheeler's original delayed-choice experiment was a thought experiment that became the basis for actual experiments. Because these experiments illustrate that our observations of past events are influenced by choices we make in the present, they reveal the existence of another kind of locality. In the delayed-choice experiments, the collapse of the wave function occurs over any distance and is insensitive to the arrow of time. The second type of nonlocality revealed in these experiments that we must also recognize as a fact of nature is what we term "temporal nonlocality."

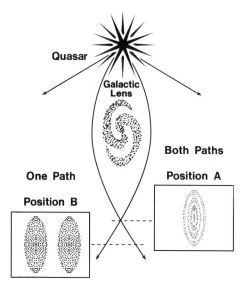

FIGURE 24. Wheeler's gravitational lens experiment uses a gravitational galactic lens to perform a "delayed-choice" experiment. Depending on where one places the light detector, one determines "now" what path or paths (option *A*: both paths; option *B*: one path) the photon "took" on its way to Earth.

To illustrate the importance of temporal nonlocality in making cosmological observations, let us consider the following ingenious delayed-choice experiment devised by Wheeler (see Fig. 24). In this experiment, light emitted from a quasar passes by an intervening galaxy that serves, in effect, as a gravitational lens. The light can be presumed to be traveling in two paths—in a straight-line path from the quasar and in a bent path caused by the gravitational lens. Inserting a half-silvered mirror at the end of the two paths with a photon detector behind each mirror, we should arrive at the conclusion that the light has indeed followed the two paths.

If, however, we observe the light in the absence of the half-silvered mirror, our conclusion would be that the light traveled in only one path and only one detector would register a photon. What we have done here is chosen to measure the wave or particle complementary aspects of this light with quite different results. The wave aspect is apparent when we insert the half-silvered mirror as reflection, refraction, and interference, and the particle aspect is apparent when the mirror is removed and light is observed by the single photon detector.

What is dramatic in this experiment is that we are determining the path of light traveling for billions of years by an act of measurement in the last nanosecond. In accordance with Bohr's Copenhagen interpretation, however, conferring reality on the photon path without taking into account the experimental setup is not allowed. This reality cannot be verified in the absence of observation or experiment. What is determined by the act of observation is a "view" of the universe in our conscious construction of the reality of the universe.

Although our views of this reality are clearly conditioned by acts of observation, the existence of the reality itself is not in question. However, because we cannot take the reality of the photon path for granted,[7] our observational "view" in the present influences our "view" of the past. It also seems clear that this situation would be ambiguous if we did not take into account the manner in which the results are conditioned by the act of observation.

What astronomers normally do is implicitly confer reality on a photon path without considering the conditions of observation. In this particular situation, factoring out the observational limits that might have been occasioned by quantum indeterminacy does not eliminate this problem. This is an epistemological problem that cannot be circumvented by any appeal to "practical" solutions. If the collapse of the wave function occurs with any quantum transition, which seems more reasonable and certainly less anthropomorphic than the idea that it occurs only as a consequence of measurement by intelligent observers, our observations of light from distant reaches of the universe may be conditioned by observational limits that can be large in any given observation.

Viewing this problem in its proper context, however, we are not driven to the conclusion that gathering observational data from light from distant reaches of the universe is useless. This is anything but the case. But it does suggest that we must re-examine the experimental situation in terms of what it clearly implies about spatial and temporal nonlocalities. In doing so, we discover that we are dealing not merely with two types of nonlocality but three. More accurately, the two complementary spatial and temporal nonlocalities imply the existence of a third type of nonlocality whose existence cannot be directly confirmed.

Three Types of Nonlocalities

Spatial, or type I nonlocality, is shown in Fig. 25, where photon entanglement persists at all levels across space-like separated regions even over cosmological scales. Temporal, or type II nonlocality, is shown in Fig. 26, where the path that a photon follows is not determined until a delayed choice, shown at the origin of the diagram, is made. Here the path of the photon is a function of the experimental choice, and this nonlocality could occur over cosmological distances.[8] Type III nonlocality, which represents the unified whole of space-time, is revealed in its complementary aspects as the unity of space (type I nonlocality) and the unity of time (type II nonlocality). Hence, type III nonlocality exists *outside* the framework of space and time.

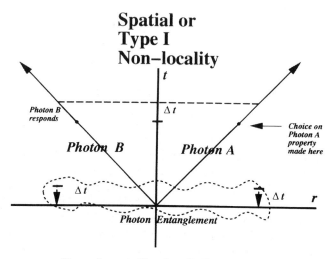

Spatial or Type I Non-locality

Experiment Testing Bell's Inequality

FIGURE 25. Type I nonlocality.

Temporal or Type II Non-locality

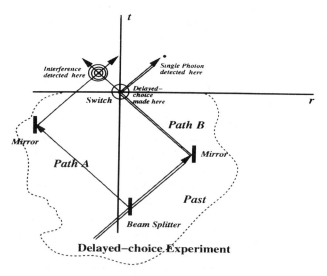

Delayed–choice Experiment

FIGURE 26. Type II nonlocality.

While types I and II taken together as complementary constructs describe the entire physical situation, neither can individually disclose this situation in any given instance. This is because the reality represented by type III nonlocality is the unified whole of space-time revealed in its complementary aspects as the unity of space (type I nonlocality) and the unity of time (type II nonlocality). Although we can confirm with experiment the existence of types I and II, which taken together imply the existence of type III, the existence of type III cannot be directly confirmed by experiment.

The third nonlocality refers to the undivided wholeness of the cosmos. And spatial and temporal nonlocalities taken together mark the event horizon where we confront the existence of this whole that cannot be directly confirmed. There are obviously spatial and temporal nonlocalities in acts of observation in astronomy, but neither can serve as the basis for developing new physical theory for the same reason that the results of the experiments proving Bell's inequality do not lead to additional theory. What is revealed in both instances are "aspects" of reality as a whole as opposed to the behavior of parts.

But, in conducting experiments, we do not "cause" the past to happen or "create" nonlocal connections. We are simply demonstrating the existence of the part-whole complementarity in our efforts to coordinate our knowledge of the parts. What comes into existence as an object of knowledge was not created or caused by us for the simple reason that it was always there—and the "it" in this instance is a universe that seems to exist on a primary level as an undivided wholeness.

How, then, do these nonlocalities serve the progress of science? One answer is that they clarify how one should view the universe and what should be known about the universe in epistemological terms. Because the three nonlocalities suggest that complementarity is a primary feature of the universe on the most fundamental level, this suggests that the future progress of scientific thought might lie in the direction of heretofore undiscovered complementary constructs.

Complementarity and Cosmology

The suggestion that future progress in science may be marked by the discovery of additional profound complementarities becomes more reasonable when we realize that this is precisely the direction in which modern physics has moved in the past. Equally interesting, these complementarities seem to be emergent at increasingly larger times and scales. To illustrate both points, let us appeal to a type of diagram that we call "universal diagrams."[9] The particular universal diagram shown in Fig. 27 is based on all the known scales in the universe that correlate with evolving, or unfolding, order in terms of complementary relations. The horizontal axis represents scale, or size of objects in the universe, ranging some sixty-one orders of magnitude from the Planck length, 10^{-33} cm, to the radius of the observable universe at about 10^{28} cm or roughly ten billion light years. The vertical axis roughly correlates emergent complementary relations with the scale at which they appear. In Fig. 28, we show a universal diagram plotting the mass versus size of various classes of objects in the universe. This diagram provides the appropriate scales to better appreciate what is represented in Fig. 27.

At the beginning of the universe, we witness the appearance of the first complementarity between the nothingness of the vacuum state and the somethingness of the quantum of action. It was, we now speculate, a fluctuation in the vacuum state caused by the quantum of action that resulted in origins. This first complementarity is then infolded into all subsequent events. Between 10^{-16} and 10^{-8} cm, the complementarities between field and quanta and particle-wave emerge and are folded in the unified process. On roughly this same scale, there also emerges out of the unified field complementary relations between the four known fields interacting in terms of the enfolded complementarities of quantum of action-vacuum, quanta-field, and particle-wave. All of these complementarities become the basis for the part-whole complementarity that is evinced at all scales.

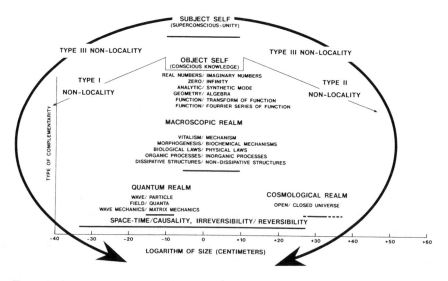

FIGURE 27. Complementarity universal diagram showing unfolding complementary relationships for different scales.

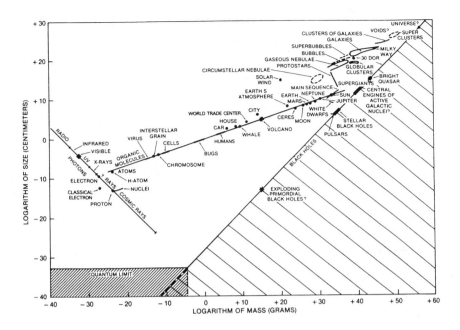

FIGURE 28. Mass versus size universal diagram for various classes of objects in the universe.

From the perspective of mathematical physics, a theory that allows us to better understand the interaction of quanta and fields at all times and scales would appear to be a "theory of everything." There is, however, one aspect of physical reality that will probably not be explained by such a theory—biological or organic life. Although we have argued that the relationship between organic and inorganic matter can be better understood based on a new part-whole complementarity that features emergence, the laws associated with distinctly biological processes appear to be very different from the laws of physics. We represent these laws, therefore, as complementary constructs.

Beginning at scales only a few thousand atoms across, or a fraction of a micron, one encounters the first viruses. These viruses display properties and behavior associated with life that cannot be fully comprehended by mathematical physics. In terms of scale, one could represent the realm of organic matter with the largest life form know to us—the blue whale. But it is probably more appropriate, given the interdependence and interconnectedness of all biological life, to represent it as the entire biosphere. Using the biosphere as the upper limit and the viruses as the lower limit, the scale from the emergence of the complementarities between organic and inorganic matter ranges from a few microns to a few thousand kilometers. Note that all the additional and related complementarities discussed earlier are also represented in Fig. 27.

To complete this picture, let us now move into the realm of the astronomical beyond our solar system. Here we discover collapsed neutron stars only a few kilometers across, stellar black holes, terrestrial planets, collapsed white dwarfs, and much larger regular "main-sequence" stars. There are also red giant and red supergiant stars that can span dimensions larger than the inner solar system. Beyond the dimensions of all stellar objects fall the aggregates of stars, star-clusters, and the gaseous nebular regions from which all stars are born—molecular regions, neutral hydrogen regions, and ionized hydrogen regions.

Moving into the galactic scale, we discover in ascending order tiny dwarf ellipticals, the small irregular galaxies, the common spiral galaxies like the Milky Way, and, finally, the most massive and largest galaxies, known as giant ellipticals. Outside of this enormous scale we encounter giant new structures composed of clusters and super-clusters of galaxies with dimensions spanning hundreds of millions of light years. As mentioned earlier, the observable universe appears to be filled with these large superclusters.

The complementarities present in the mathematical theory (upper part of Fig. 27) that allow us to coordinate such an impressive range of experience are obviously without scale. The discovery that these constructs are not pre-existing and disembodied forms has not been, as we have seen throughout, entirely welcomed by the mathematical physicist. If, however, we accept the conclusion that complementarities are foundational to our conscious constructions of reality in mathematical language, and if we factor in the conclusion that all events in the cosmos are folded in this seamless unity, the complementarities in mathematical theory can be viewed in a very different way.

Perhaps the reason the description of physical reality in mathematical physics allows us to coordinate experience with this reality is that the logical principle of complementarity describes a fundamental dynamic in all of our conscious constructions or representations of reality. Since it seems clear that the logical framework of the complementarity describes primary oppositions in mathematical language, this could explain why this language is "privileged" in our efforts to coordinate experience with physical reality. The old view that mathematical language is privileged because it allows the human mind to commune with pre-existent immaterial forms no longer seems supportable. Perhaps the more appropriate and logically consistent view is that the language of mathematical physics is privileged in its ability to coordinate experience with physical reality because the logical principle of complementarity is the fundamental structuring principle in this reality.

This hypothesis is made even more reasonable when we consider that complementarity appears to have manifested as a macro-level phenomenon in the evolution of life on Earth beginning some four billion years ago. Since the human brain is the most complex expression of the process of evolution to date, the emergence of complementarity as a fundamental dynamic in the creation of complex symbol systems, or conscious constructions of reality, could be a higher-level manifestation of complementarity in the process of evolution. In the absence, however, of any linkage between quantum mechanical processes in the human brain invoking complementarity and global brain function, this is merely a hypothesis that might be worthy of further investigation.

In the next chapter we will demonstrate that the hidden metaphysical presupposition that is foundational to a belief in a one-to-one correspondence between every element of the physical theory and physical reality has continued to work its magic on theoretical physicists. In an effort to "save" this correspondence, a number of physicists have posited theories with large cosmological implications that attempt to subvert wave-particle dualism and Bohr's view of the quantum measurement problem. We will attempt to show that these physicists have made "metaphysical leaps" in the service of the hidden ontology of classical epistemology and that Bell's theorem and the experiments testing that theorem clearly reveal why this is the case.

8
Quantum Ontologies: Metaphysics in Modern Physics

...May God us keep
From single Vision and Newton's sleep!

William Blake

The insurmountable problem in preserving the classical view of correspondence in the face of the evidence disclosed by Bell's theorem and the recent experiments testing that theorem has been defined by physicist Henry Stapp. The simultaneous correlations of results between space-like separated regions in the Aspect experiments indicate that nonlocality is a fact of nature. However, one cannot posit any causal connection between these regions in the absence of faster-than-light communication. As Stapp put it:

> No metaphysics not involving faster-than-light propagation of influences has been proposed that can account for all of the predictions of quantum mechanics, except for the so-called many-worlds interpretation, which is objectionable on other grounds. Since quantum physicists are generally reluctant to accept the idea that there are faster-than-light influences, they are left with no metaphysics to promulgate.[1]

If light speed is the ultimate limit at which energy transfers or signals can travel, and if any attempt to measure or observe involves us and our measuring instruments as integral parts of the experimental situation, we are forced to conclude that the correlations evident in the Aspect and Gisin experiments can be explained only in terms of a strange fact: the system, which includes the experimental setup, is an unanalyzable whole. When we consider that the universe has also been evolving since the big bang via the exchange of quanta in and between fields, the fact that nonlocality has always been a feature of this process leads to other more formidable conclusions. Since all quanta have interacted with one another in a single quantum state and there is no limit to the number of particles that can interact in a single quantum state, the universe on a very basic level could be a "single" quantum system that responds together for further interactions.

If nonlocality is an indisputable fact of nature, indeterminacy is also an indisputable fact of nature. The only way to retain belief in the classical view of correspondence is to presume the existence of that which cannot be proven by theory or experimental evidence—faster-than-light communication. The central question in this chapter is whether the one-to-one correspondence between every element of the physical theory and physical reality is possible in this situation—or anywhere else in the quantum domain. If it is possible, we can presume that there is a viable alternative to Bohr's Copenhagen Interpretation and therefore that the mathematical description of nature as Einstein conceived it could be sustained. But attempts to preserve this view not only require metaphysical leaps that result in unacceptable levels of ambiguity. They also fail to meet the requirement that testability is required to confirm the validity of any physical theory.

The Quest for a New Ontology

According to Henry Stapp, the three "principal ontologies that have been proposed by quantum physicists" as alternatives to Bohr's Copenhagen Interpretation are the "pilot-wave ontology" of de Broglie and Bohm, the "many-worlds interpretation" of Everett, Wheeler, and Graham, and the "actual event ontology."[2] Although the actual event ontology is most closely associated with Heisenberg, it also proceeds along lines of argumentation suggested by Bohm and Whitehead. The following is a summary of Stapp's more detailed commentary on each of these ontologies.[3]

The pilot-wave model ontologizes, or confers an independent and unverifiable existence on, what is termed the "quantum potential," and it is based on David Bohm's notion that the universe is an unbroken wholeness and that parts "manifest" from this whole. This wholeness, which Bohm termed the "implicate order," is described as an unbroken web of cosmic interconnectedness.[4]

In the pilot-wave ontology, a nonrelativistic universe is described in terms of the square of the absolute value of the wave function P and its phase S, and the wave function is completely defined by the quantities S and P. The quantity P serves the same function as the square of the absolute value of the wave function does in orthodox quantum theory; it defines the probability that the particle will be found within a given region. The phase S is called the "quantum potential." The phase of a wave gives essential information about the way a wave should be added to another wave. This addition of waves, as we have seen, is a central feature of all wave phenomena, including quantum superposition phenomena.

The central feature of this ontology is the assumption that the quantum potential is a mathematical function that fills all space-time in the implicate order and exists, in some sense, beyond space-time. Bohm seeks to justify the view that the quantum potential exists in the implicate order with the following argument. Because the quantum potential, like all phases of waves, is not directly observable, it is assumed to exist underneath the space-time level of all quantum phenomena at a subquantum level.

In this ontology, a velocity field is defined by the rate that the quantum potential S changes in space. Because the mathematical functions, P and S, are considered sufficient to generate the individual particle trajectories, it is assumed that definite trajectories can be retained in spite of the quantum measurement problem and that physical reality is completely deterministic in the classical sense. All trajectories of particles are classical space-time trajectories, and the mathematical theory is presumed to correspond with all aspects of this reality. Because the trajectory of each particle is derived by the underlying quantum potential, this would also appear to provide an explanation for quantum nonlocality.[5] If the quantum potential exists in all space and time, the correlated results in the Aspect experiments could result from interconnections within the system at a deeper level in apparent defiance of the finite speed of light.

One large problem with the pilot-wave model is that it says nothing about the initial conditions that must be specified to determine the quantum potential. Moreover, the model does not explain why some possibilities given by the wave function are realized when an observation is made and others are not. As Stapp notes, this problem is "bypassed" by assuming that the other branches of the wave function are "empty" and have no influence on anything physical.

Although the model seeks to reconstruct the classical correspondence between physical theory and reality, "only" the probability P is testable in the laboratory. S, in contrast, is completely unobservable. Because the model ontologizes, or confers an independent and untestable existence on, the quantum potential S, it clearly violates the well-worn scientific precept that any predictions of physical theory must be subject to experimental proof.

In the many-worlds interpretation, the wave function is ontologized, or presumed to have an independent and unverifiable existence, in a more radical sense. Here the fundamental reality in the universe "is" the wave function, and nothing else need be taken into account except for the consciousness of human observers. As a measurement is made by a human observer, all possibilities described by the wave function "must be" realized for the simple reason that the wave function is assumed to be "real."

When an observation is made in this model, all of the mathematically real possibilities given by the wave equation are allegedly realized, and there are no empty branches. The assumption is that some of these "real" possibilities are actualized by an observer in one world and the other "real" possibilities are actualized by an observer in another world. According to this ontology, the room in which you now sit is splitting into virtually identical rooms with virtually identical observers billions of times per second. And yet any single observer is not aware that this multitude of different universes is perpetually coming into existence because all of the real possibilities in the wave function cannot be realized in a single act of observation. Here again, the decision to ontologize the wave function takes us out of the realm of experimental physics and there is simply no way to prove that the other worlds exist. Hence the impulse to preserve complete correspondence between physical theory and physical reality in the many-worlds interpretation obviates any opportunity to confirm that correspondence in experiment.

Another large problem with the many-worlds interpretation concerns initial conditions. If all branches of the wave equation are "ontologically equivalent" and the universe is a mixture of all possible conditions given by the equation, how are initial conditions established? Put another way, how could anything actual emerge from something so amorphous?[6]

If one assumes that the physical system has already separated into discrete branches, one could presume that the element of discreteness has already been introduced into the observed system.[7] If, however, we view the wave function as a continuous superposition of all macroscopic possibilities, the result is an amorphous superposition of a continuum of different states. Because this translates mathematically into zero probability, the existence of a conscious observer registering specific measurements in quantum mechanical experiments is quite improbable. It is also clear that an "economical" description in this instance does not result in greater economy when we consider the vast number of parallel universes that results. If nature tends to be economical, this tendency is clearly violated in the many-worlds interpretation.

In the actual-events ontology proposed by Werner Heisenberg, the fundamental process of nature is viewed as a sequence of discrete actual events. In this view, the "potentialities" created by a prior event become the "potentialities" for the next event. The discontinuous change of the wave function is viewed here as describing the "probability" of an event that becomes an "actual" or "real" event when the measuring device acts on the physical system.[8]

The assumption is that the discontinuous change in our knowledge at the moment of measurement is equivalent to the discontinuous change of the probability function. The real, or actual, event is represented by the quantum jump in the absolute wave function. Thus the probability amplitude of the absolute wave function corresponds with the "potentia," or the objective tendency to occur, as an actual event, and is disassociated from the actual event.

What is ontologized in the actual-events ontology is an alleged aspect of the wave function, the quantum potential, that is somehow empowered to "select" or "choose" a particular macroscopic variable before the act of observation. The model does not provide a detailed mathematical description of how this transition from possible to actual occurs and does not allow for any experimental proof of the existence of the quantum potential. Hence this model, like the others, is not subject to experimental proof and must be viewed in scientific terms as ad hoc and arbitrary.

Stapp has proposed his own version of the actual event ontology. Although he concedes that Bohr's CI must be invoked to understand quantum mechanical events that are not observed, or that occur "outside" the human brain, he claims that the wave function collapses into single high-level classical branches, rather than lower-level states, "within" the human brain. The obvious question here is: Why does the quantum reality exist as such "outside" the human brain and become classical "inside" the human brain?

Stapp's answer is that "evolutionary pressures" on our species were such that they tended to push collapses to higher levels. In other words, ancestors who perceived the collapses in what would eventually be described as classical terms had a survival advantage and were more likely to pass on this trait to their offspring. Obviously, this alleged transformation in the manner in which the quantum potential was recorded in the human brain during the course of evolution is not subject to verification in controlled and repeatable experiments and must be viewed as little more than philosophical speculation.[9]

Another related argument has been advanced by the mathematician Roger Penrose. Penrose claims that epistemological problems in quantum physics could be resolved by some future theory of quantum gravity that features noncomputable elements. In an attempt to provide a link between this unknown theory of quantum gravity and neurons in the human brain, he draws on Stuart Hameroff's studies on microtubules.[10] Most neuroscientists agree that microtubules, which provide a skeleton for the neuron, control the shape of the neuron and serve to transport molecules between the cell body and synapses. Penrose goes beyond this consensus and speculates that the network of microtubules acting "in concert" in the human brain could serve another function. They could, he argued, collapse the wave function, and this could result in the noncomputability that he believes is necessary for human consciousness.[11]

The basic argument advanced by Penrose has been widely criticized[12] *and we will not review that criticism here. What is most interesting for our purposes is that this attempt to "ground" consciousness in a quantum mechanical process privileges the collapse of the wave function. This particular violation of the assumption that wave and particle are complementary aspects of the total reality resembles the quantum ontologies discussed here in two respects. It makes the foundations of consciousness more ambiguous and grandly oversimplifies the complexities of the physical situation. Yet Penrose's assumption that the actual dynamics of consciousness are not computable or reducible to a set of algorithms has merit and should continue to be explored.*

In all of these examples, the decision to ontologize, or to confer an independent and unverifiable existence on, the wave function or some aspect of the function disallows the prospect of presenting any new physical content that can be verified under experimental conditions. It seems clear that the impulse here is not to extend the mathematical description to increasingly greater verifiable limits. It is to sustain the classical view of one-to-one correspondence between every element of the physical theory and physical reality.

If, however, we practice epistemological realism and refuse to make metaphysical leaps, wave and particle aspects of quantum reality must be viewed as complementary. Neither aspect constitutes a complete view of this reality, both are required for a complete understanding of the situation, and observer and observed system are inextricably interconnected in the act of measurement and in the analysis of results. Hence there is no one-to-one correspondence between the physical theory and physical reality.

If we ignore the limitations inherent in observation and measurement occasioned by the existence of the quantum of action and seek to affirm this correspondence in the absence of experimental evidence, this not only represents a violation of scientific method; it also obliges us to make a metaphysical leap by ontologizing one aspect of quantum reality. This logical mistake results, as Bohr said it would, in ambiguity, and it carries the totally unacceptable implication that metaphysics is prior to physics.

The New Epistemology in a Philosophical Context

All scientific truths, as Schrödinger said, "are meaningless outside their cultural context," and the classical view of correspondence was a product of that context. As we have seen, the received logical framework for arriving at truth in rational discourse coupled with belief in metaphysical or ontological dualism gave birth to the ontology of classical epistemology. The success of the classical paradigm coupled with the triumph of positivism in the nineteenth century served to disguise the continued reliance on seventeenth-century presuppositions in the actual practice of physics.

For all the reasons mentioned earlier, the experiments testing Bell's theorem now force us to abandon the classical view of correspondence and the related idea that mathematical forms and ideas have an independent or separate existence in physical reality. Because a one-to-one correspondence between all aspects of the physical theory and the physical reality does not exist in a quantum mechanical universe, we must now view the truths of physical theory in the manner advocated by Bohr. Although physical theory has served to coordinate our experience with nature beautifully, we can on longer regard the truths revealed by these theories as having an independent existence. These truths, like other truths, exist in our world-constructing minds.

This does not mean, as we have continually stressed, that the authority of scientific knowledge is diminished or compromised in the least. For a scientific construct to be recognized and perpetuated as such, it must continually stand before the court of last resort—repeatable experiments under controlled conditions. And that court, as we saw in the experiments testing Bell's inequality, will not modify its verdict based on any special pleading about the character of any defendant.

The primary source of our confusion in analyzing the results of the experiments testing Bell's inequality is that we have committed what Whitehead termed the "fallacy of misplaced concreteness." We have accepted abstract theoretical statements about concrete material results in terms of single categories and limited points of view as totally explanatory. The fallacy is particularly obvious in our dealings with the results of the Aspect and Gisin experiments. Although the results "infer" wholeness in the sense that they show that the conditions for this experiment constitute an unanalyzable and undissectable whole, the abstract theory that helps us coordinate the results cannot "in principle" disclose this wholeness. Because the abstract theory can deal only in complementary aspects of the complete reality disclosed in the act of measurement, that reality is not itself—in fact or in principle—disclosed.

Uncovering and defining the whole in mathematical physics did seem realizable prior to quantum physics because classical theory was presumed to mirror exactly the concrete physical reality. An equally important and essential ingredient in the realization of that goal was the belief in classical locality, or in the essential distinctness and separability of space-like separated regions. Because classical epistemology and the assumption of locality allowed one to presume that the whole could be described as the sum of its parts, it was assumed that the ultimate extension of theory to a description of all the parts would disclose the whole. With the discovery of nonlocality, it seems clear that the whole is not identical to the sum of its parts and that no collection of parts, no matter how arbitrarily large, can fully disclose or define the whole. As we saw earlier, this also appears to be case for the whole of the biota in biological reality.

Parallels with Eastern Metaphysics

In this discussion of physics and metaphysics, we should probably say something here about the alleged parallels between the "holistic" vision of physical reality in modern physics and religious traditions featuring holism, or ontological monism, like Hinduism, Taoism, and Buddhism. The extent to which the study of modern physical theories can entice one to embrace the eastern metaphysical tradition is nicely illustrated in an interview with David Bohm. In this interview, Bohm commented that, "Consciousness is unfolded in each individual," and meaning "is the bridge between consciousness and matter." Other assertions in the same interview, like "meaning is being," "all moments are one," and "now is eternity" would be familiar to anyone who has studied eastern metaphysics.[13]

Eastern philosophies can be viewed on the level of personal belief or conviction as more parallel with the holistic vision of nature featured in modern physical theory. It is, however, impossible to conclude that eastern metaphysics legitimates modern physics or that modern physics legitimates eastern metaphysics. The obvious reason for this is that orthodox quantum theory, which remains unchallenged in its epistemological statements, disallows any ontology. In addition, the recent discovery that nonlocality is a fact of nature does nothing to change this situation. Although this discovery may imply that the universe is holistic, physics can say nothing about the actual character of this whole.

If the universe were, for example, completely described by the wave function, this need not be the case. One could then conclude that the ultimate character of the whole, in its physical analog at least, had been "revealed" in the wave function. We could then assume that any sense we have of profound unity or mystical oneness with the cosmos has a direct analog in physical reality. In other words, this experience of unity with the cosmos could be presumed to correlate with the action of the deterministic wave function that governs the locations of particles in our brain and the direction in which they are moving. From this perspective, the results of the Aspect and Gisin experiments could be providing a kind of scientific proof for ontological monism.

The problem in quantum theory, however, is that the wave function only provides "clues" about possibilities of events rather than definite predictions of events. But what if we assume that the sense of unity with the whole is associated with some integral or integrated property of the wave function of the brain and that of the entire universe? The problem here is the same as that associated with the many-worlds interpretation; there is simply no way in the physical theory for a discrete experience of unity to emerge from the incoherently added wave functions corresponding with the multitude of quanta in a human brain.

Although our new epistemological situation suggests that questions regarding the character of the whole no longer lie within the domain of science, this does not prevent us from exploring the implications in philosophical terms. However, if we are to properly understand these implications, it is first necessary to know more about the origins and history of classical epistemology and the associated doctrine of correspondence. We will demonstrate that what we term the "hidden ontology of classical epistemology" has been an overt and covert presence in the practice of physics since the seventeenth century. We will also argue that the experiments testing Bell's theorem now oblige us to purge this last vestige of metaphysics from physics.

The fact that these experiments can only infer, as opposed to prove, the existence of the indivisible whole indicates that the classical ambition to explain the whole in terms of the sum of its parts cannot be fulfilled. Whether one chooses to view this whole as having a metaphysical dimension is, for reasons that should soon become obvious, a matter of personal choice and conviction. However, it also seems clear that because this whole cannot be reduced to or described by physical theory, the idea, which originated in the eighteenth century, that the scientific description of nature does not allow for the presence of any extra-scientific agency or principle is no longer valid. If this is this case, we may be entering a new phase of human consciousness, where science will no longer seem hostile to profound religious feeling and where we will witness a renewed dialog between these disparate ways of knowing.

9
The Ceremony of Innocence: Physics, Metaphysics, and the Dialog between Science and Religion

> I will not go so far as to say that to construct a history of thought without profound study of the mathematical ideas of successive epochs is like omitting Hamlet from the play that is named after him....But it is certainly analogous to cutting out the part of Ophelia.
>
> *Alfred North Whitehead*

In the history of science, there are many ironies, but perhaps the greatest of these concern the origins, evolution, and ultimate demise of classical epistemology and the doctrine of positivism. For reasons that should soon become obvious, belief in the seventeenth presupposition of metaphysical dualism, more formally known as ontological dualism, was critically important during the first scientific revolution and there are good reasons to believe that this revolution may not have occurred in its absence.

During the eighteenth and nineteenth centuries, attempts to purge physics of all metaphysical and nonmathematical constructs resulted, as we have seen, in the doctrine of positivism. As it turned out, however, the epistemology that this doctrine was designed to protect was premised on an unexamined article of faith—that physical theory is a logically consistent and self-referential system that could disclose with complete certainty the essence of physical reality.

The first challenges to this view came from pure, as opposed to applied, mathematics. The invention of non-Euclidian geometries threatened the classical view of the structure of space and its relation to time. Failed attempts, from Cantor to Gödel, to prove that mathematics is a self-referential system that rested on firm logical foundations strongly indicated that the classical view of correspondence was flawed. Within the normal conduct of physics, the discovery of wave-particle dualism and quantum indeterminacy posed the first major threat to the efficacy of classical epistemology and the associated doctrine of positivism. As noted earlier, Einstein's attempt to preserve and protect this epistemology culminated in the EPR thought experiment that enticed John Bell to develop the theorem that became the basis for actual experiments. It was these experiments that eventually decided the Einstein-Bohr debate in favor of Bohr and provided a conceptual basis for the new epistemology of science that we have attempted to articulate better here.

One irony is that the physics that had allegedly purged itself of all metaphysical constructs was premised on what we have termed here the "hidden ontology of classical epistemology." Hence the progress of this physics was deeply wedded to a metaphysical quest. The quest was to subsume all of physical reality with physical theory and to demonstrate that all its constituent parts could be assembled in one coherent whole. However, because the progress of science is tightly constrained by physical theory and experimental evidence, meaning in this case Bell's theorem and the experiments testing the predictions made in the theorem, we are now obliged to recognize the existence of this metaphysical quest.

When the results of these experiments are properly analyzed in the absence of metaphysical assumptions, this not only forces us to abandon classical epistemology and the doctrine of correspondence but also reveals that the whole whose existence is inferred in experiments testing Bell's theorem cannot be fully disclosed or described by physical theory and that the parts exist in some sense within this whole. The final irony is, therefore, that after we finally purged physics of metaphysical constructs, we are confronted with a fundamental reality that exists completely outside the domain of physics. To complete this picture and better explore its philosophical implications, we will briefly consider the origins and history of the hidden ontology of classical epistemology.

Physics and Seventeenth-Century Metaphysics

The most fundamental aspect of the western intellectual tradition is the assumption that there is a division between the material and immaterial worlds or between the realm of matter and the realm of pure mind or spirit. The metaphysical framework based on this assumption is known as "ontological dualism." As the word "dual" implies, the framework is predicated on an ontology, or a conception of the nature of God or Being, within which reality is assumed to have two distinct and separable dimensions.

The notion that the material world experienced by the senses is inferior to the immaterial world experienced by mind or spirit has been blamed for frustrating the progress of physics to at least the time of Galileo. In one very important respect, however, it made the first scientific revolution possible. Copernicus, Galileo, Kepler, and Newton firmly believed that the immaterial geometrical and mathematical ideas that "inform" or "give form to" physical reality had a prior existence in the mind of God and that doing physics was a form of communion with these ideas. It was this belief that allowed these figures to assume that geometrical and mathematical ideas could serve as transcriptions of the actual character of physical reality and could be used to predict the future of physical systems.

Copernicus would have been described by his contemporaries as an administrator, diplomat, avid student of economics and classical literature, and, most notably, a highly honored and placed church dignitary. Although we named a revolution after him, this devoutly conservative man did not set out to create one. The placement of the sun at the center of the universe seemed right and necessary to Copernicus not as a result of making careful astronomical observations; he, in fact, made very few of these observations while developing his theory, and then only to ascertain if his prior conclusions seemed correct. The Copernican system was not any more useful in making astronomical calculations than the accepted model, and in some ways it was much more difficult to implement. What, then, was his motivation for creating the model and his reasons for presuming that the model was correct?

Copernicus felt that the placement of the sun at the center of the universe made sense because he viewed the sun as the symbol of the presence of a supremely intelligent and intelligible God in a man-centered world. He was apparently led to this conclusion in part because the Pythagoreans believed that "fire" exists at the center of the cosmos, and Copernicus identified this fire with the fireball of the sun. The only support Copernicus could offer for the greater efficacy of his model is that it represented a simpler and more mathematically harmonious model of the sort that the Creator would obviously prefer. The language used by Copernicus in *The Revolution of Heavenly Orbs* illustrated the religious dimension of his scientific thought: "In the midst of all the sun reposes, unmoving. Who, indeed, in this most beautiful temple would place the light-giver in any other part than whence it can illumine all other parts?"[1]

The belief that the mind of God as Divine Architect permeates the workings of nature was the guiding principle of the scientific thought of Johannes Kepler. For this reason, most modern physicists would probably feel some discomfort reading Kepler's original manuscripts. Physics and metaphysics, astronomy and astrology, geometry and theology commingle with an intensity that might offend those who practice science in the modern sense of that word. Physical laws, wrote Kepler, "lie within the power of understanding of the human mind; God wanted us to perceive them when he created us in His image in order that we may take part in His own thoughts....Our knowledge of numbers and quantities is the same as that of God's, at least insofar as we can understand something of it in this mortal life."[2]

Believing, like Newton after him, in the literal truth of the words of the Bible, Kepler concluded that the word of God was also transcribed in the immediacy of observable nature. Kepler's discovery that the motions of the planets around the sun were elliptical, as opposed to perfect circles, may have made the universe seem a less-perfect creation of God in ordinary language. For Kepler, however, the new model placed the sun, which he also viewed as the emblem of divine agency, more at the center of a mathematically harmonious universe than the Copernican system allowed. Communing with the perfect mind of God required, as Kepler put it, "knowledge of numbers and quantity."

Because Galileo did not use, or even refer to, the planetary laws of Kepler when those laws would have made his defense of the heliocentric universe more credible, his attachment to the god-like circle was probably a more deeply rooted aesthetic and religious ideal. But it was Galileo, even more than Newton, who was responsible for formulating the scientific idealism we are now forced to abandon. In *Dialog Concerning the Two Great Systems of the World*, Galileo said the following about the followers of Pythagoras: "I know perfectly well that the Pythagoreans had the highest esteem for the science of number and that Plato himself admired the human intellect and believed that it participates in divinity solely because it is able to understand the nature of numbers. And I myself am inclined to make the same judgment."[3]

This article of faith—that mathematical and geometrical ideas mirror precisely the essences of physical reality—was the basis for the first scientific revolution. Galileo's faith is illustrated by the fact that the first mathematical law of this new science, a constant describing the acceleration of bodies in free fall, could not be confirmed by experiment. The experiments conducted by Galileo in which balls of different sizes and weights were rolled simultaneously down an inclined plane did not, as he frankly admitted, yield precise results. However, because vacuum pumps had not yet been invented, there was no way Galileo could subject his law to rigorous experimental proof in the seventeenth century. Galileo believed in the absolute validity of this law in the absence of experimental proof because he believed that movement could be subjected absolutely to the law of number. What Galileo asserted, as the

French historian of science Alexander Koyre put it, is "that the real is in its essence, geometrical and, consequently, subject to rigorous determination and measurement."[4]

The popular image of Isaac Newton is that of a supremely rational and dispassionate empirical thinker. Newton, like Einstein, had the ability to concentrate unswervingly on complex theoretical problems until they yielded a solution. But what most consumed his restless intellect was not the laws of physics. In addition to believing, like Galileo, that the essences of physical reality could be read in the language of mathematics, Newton also believed, with perhaps even greater intensity than Kepler, in the literal truths of the Bible.

For Newton the mathematical language of physics and the language of biblical literature were equally valid sources of communion with the eternal and immutable truths existing in the mind of God. Newton's theological writings in the extant documents alone consist of more than a million words in his own hand, and some of his speculations seem quite bizarre by contemporary standards. Earth, said Newton, will still be inhabited after the day of judgment, and heaven, or the New Jerusalem, must be large enough to accommodate both the quick and the dead. Newton then put his mathematical genius to work and determined the dimensions required to house this population. His rather precise estimate was "the cube root of 12,000 furlongs."

The point is that during the first scientific revolution the marriage between mathematical idea and physical reality, or between mind and nature via mathematical theory, was viewed as a sacred union. In our more secular age, the correspondence takes on the appearance of an unexamined article of faith or, to borrow a phrase from William James, "an altar to an unknown god." Heinrich Hertz, the famous nineteenth-century German physicist, nicely described what about the practice of physics tends to inculcate this belief: "One cannot escape the feeling that these mathematical formulae have an independent existence and intelligence of their own, that they are wiser than we, wiser than their discoverers, that we get more out of them than was originally put into them."[5]

Although Hertz made this statement without having to contend with the implications of quantum mechanics, the feeling he described remains the most enticing and exciting aspect of physics. That "elegant" mathematical formulas provide a framework for understanding the origins and transformations of a cosmos of enormous age and dimensions is a staggering discovery for budding physicists. Professors of physics do not, of course, tell their students that the study of physical laws is an act of communion with the perfect mind of God or that these laws have an independent existence outside of the minds that discover them. The business of becoming a physicist typically begins, however, with the study of classical or Newtonian dynamics, and this training provides considerable covert reinforcement of the feeling Hertz describes.

Metaphysics and Classical Physics

The role of seventeenth-century metaphysics is also apparent in metaphysical presuppositions about matter described by classical equations of motion. These presuppositions can be briefly defined as follows: (1) the physical world is made up of inert and changeless matter and this matter changes only in terms of location in space; (2) the behavior of matter mirrors physical theory and is inherently mathematical; (3) matter as the unchanging unit of physical reality can be exhaustively understood by mechanics, or by the applied mathematics of motion; and (4) the mind of the observer is separate from the observed system of matter, and the ontological bridge between the two is physical law and theory.[6] These presuppositions have a metaphysical basis because they are required to assume the following: that the full and certain truths about the physical world are revealed in a mathematical structure governed by physical laws that have a prior or separate existence from this world. It was this assumption that allowed the truths of mathematical physics to be regarded as having a separate and immutable existence outside the world of change.

As overt appeal to metaphysics became unfashionable, the science of mechanics was increasingly regarded, according to Ivor Leclerc, as "an autonomous science," and any alleged role of God as "deus ex machina."[7] At the beginning of the nineteenth century, Laplace, and several other great French mathematicians, advanced the view that the science of mechanics constituted a "complete" view of nature. Because this science, by observing its epistemology, had revealed itself to be the "fundamental" science, the hypothesis of God was, they concluded, unnecessary.

Laplace is recognized not only for eliminating the theological component of classical physics but the "entire metaphysical component" as well.[8] The epistemology of science requires, he said, that we proceed by inductive generalizations from observed facts to hypotheses that are "tested by observed conformity of the phenomena."[9] What was unique about Laplace's view of hypotheses was his insistence that we cannot attribute reality to them. Although concepts like force, mass, motion, cause, and laws are obviously present in classical physics, they exist in Laplace's view only as "quantities." Physics is concerned, he argued, with quantities we associate as a matter of convenience with concepts, and the "truths" about nature are only the quantities.

As this view of hypotheses and the truths of nature as quantities was extended in the nineteenth century to a mathematical description of phenomena like heat, light, electricity, and magnetism, Laplace's assumptions about the actual character of scientific truths seemed correct. This progress suggested that if we could remove all thoughts about the "nature of" or the "source of" phenomena, the pursuit of strictly quantitative concepts would bring us to a complete description of all aspects of physical reality. It was this impulse that led the positivists to develop a program for the study of nature that was very different from that of the original creators of classical physics.[10]

The seventeenth-century view of physics as a "philosophy of nature" or as "natural philosophy" was displaced by the view of physics as an autonomous science that was "the science of nature."[11] This view promised to subsume all of nature with a mathematical analysis of entities in motion and claimed that the "true" understanding of nature was revealed only in the mathematical description. In the history of science, as we noted earlier, the irony is that positivism, which was intended to banish metaphysical concerns from the domain of science, served to perpetuate a seventeenth-century metaphysical assumption about the relationship between physical reality and physical theory.

Einstein and the Positivists

Einstein was very much aware of the late nineteenth-century crisis over the epistemological foundations of number and arithmetic, and some of the philosophical considerations of the positivists served him well in the formulation of new physical theories. The two positivist thinkers who proved most influential on the work of Einstein were the inventor of positivism, Auguste Comte, and a contemporary of Comte mentioned earlier, Karl Weierstrass.

Comte was greatly troubled by Immanuel Kant's claim in *Critique of Pure Reason* (1781) that mathematics was essentially intuitive and that mathematical physics was less a description of the actual character of physical reality than a reflection of the ways in which humans think. Beginning in the 1850s, Comte argued that experimental science is the model for the acquisition of all knowledge, and he claimed that any knowledge that cannot be grounded in experimental proof was either theological or metaphysical. The simplest and most objective aspect of the scientific description of nature, he said, was the "mechanics of a material point." Based on this assumption, Comte concluded that the objectivity of science was grounded in a one-to-one correspondence between this "point" and the mathematics of physical theory. Because this view of the objectivity of science required an objective observer, he also concluded that this observer existed outside of or separate from the observed material world.

Weierstrass, who became the leader of positivism in mathematics, argued that mathematics could be reduced to arithmetic and sought to make the definitions and proofs of mathematics logically consistent and self-referential by reducing the number of ideal, or undefined, objects to a minimum. One such minimum, for example, was whole numbers. But Weierstrass also affirmed, in contrast with Comte, the relevance of the Kantian view by concluding that mathematics is "a pure creation of the human mind."[12]

Since Einstein was aware that attempts to posit a logically consistent foundation for number had failed, he identified with Weierstrass' view of mathematics. Yet Einstein also remained faithful to Comte's view of the objective character of scientific knowledge which, as we saw earlier, is critically dependent on two assumptions: There must be a one-to-one correspondence between every element of the physical theory and physical reality, and the observer must be separate and distinct from the observed physical system.

But as Ivor Leclerc explained, Einstein's view was not as simple as others imagined, and it contained some fundamental ambiguities.[13] Einstein may have been in full agreement with the notion that physical theories are the free invention of the human mind, but he also maintained that "the empirical contents of their mutual relations must find their representations in the conclusions of the theory."[14] Einstein sought to reconcile the fundamental ambiguity between the two positions—that physical theories "represent" empirical facts and that physical theories are a "free invention" of the human intellect— with an article of faith. "I am convinced," wrote Einstein, "that we can discover by means of purely mathematical constructions the concepts and laws connecting them with each other, which furnish the key to understanding natural phenomena."[15]

Since the lack of a one-to-one correspondence between every element of the physical theory and physical reality in quantum physics completely undermines this conviction, how does Einstein sustain it? He does so, according to Leclerc, by appealing to "a tacit seventeenth-century presupposition of metaphysical dualism and a doctrine of the world as mathematical structure completely knowable by mathematics."[16]

Einstein's View

Perhaps the best way to understand the legacy of the hidden ontology of classical epistemology in the debate over the epistemological implications of nonlocality is to examine the source of Einstein's attachment to classical epistemology in more personal terms. Einstein apparently lost faith in the God portrayed in Biblical literature in early adolescence, but as the following passage from "Autobiographical Notes" suggests, there were aspects of that heritage that carried over into his understanding of the foundations for scientific knowledge:

> Thus I came—despite the fact that I was the son of entirely irreligious (Jewish) parents—to a deep religiosity, which, however, found an abrupt end at the age of 12. Through the reading of popular scientific books I soon reached the conviction that much in the stories of the Bible could not be true. The consequence was a positively frantic (orgy) or freethinking coupled with the impression that youth is intentionally being deceived by the state through lies; it was a crushing impression. Suspicion against every kind of authority grew out of this experience.... It was clear to me that the religious paradise of youth, which was thus lost, was a first attempt to free myself from the chains of the "merely personal.".... The mental grasp of this extra-personal world within the frame of the given possibilities swam as highest aim half consciously and half unconsciously before the mind's eye."[17]

It was, suggested Einstein, belief in the word of God as it is revealed in biblical literature that allowed him to dwell in a "religious paradise of youth" and to shield himself from the harsh realities of social and political life. In an effort to recover the inner sense of security that was lost after exposure to scientific knowledge, or to become free once again of the "merely personal," he committed himself to understanding the "extra-personal world within the frame of given possibilities," or, as seems obvious, to the study of physics. Although the existence of God as described in the Bible may have been in doubt, the qualities of mind that the architects of classical physics associated with this God were not. This is clear in following comment by Einstein on the uses of mathematics:

> Nature is the realization of the simplest conceivable mathematical ideas. I am convinced that we can discover, by means of purely mathematical constructions, those concepts and those lawful connections between them which furnish the key to the understanding of natural

> phenomena. Experience remains, of course, the sole crite-
> ria of physical utility of a mathematical construction. But
> the creative principle resides in mathematics. In a certain
> sense, therefore, I hold it true that pure thought can grasp
> reality, as the ancients dreamed.[18]

This article of faith, first articulated by Kepler, that "nature is the
realization of the simplest conceivable mathematical ideas," allowed
Einstein to posit the first major law of modern physics much as it allowed Galileo to posit the first major law of classical physics.

During the period when the special and then the general theories
of relativity had not been confirmed by experiment and many established physicists viewed them as at least minor heresies, Einstein
remained entirely confident in their predictions. Ilse Rosenthal-Schneider, who visited Einstein shortly after Eddington's eclipse
expedition confirmed a prediction of the general theory (1919), described Einstein's response to this news:

> When I was giving expression to my joy that the results
> coincided with his calculations, he said quite unmoved,
> "But I knew the theory is correct," and when I asked, what
> if there had been no confirmation of his prediction, he
> countered: "Then I would have been sorry for the dear
> Lord—the theory is correct."[19]

Einstein was not given to making sarcastic or sardonic comments,
particularly on matters of religion. These unguarded responses testify
to his profound conviction that the language of mathematics allows
the human mind access to immaterial and immutable truths existing
outside of the mind that conceives them. Although Einstein's belief
was far more secular than Galileo's, it retained the same essential
ingredients.

As we have seen, what was at stake in the twenty-three-year-long
debate between Einstein and Bohr was primarily the status of an article of faith as opposed to the merits or limits of a physical theory. At
the heart of this debate was the fundamental question, "What is the
relationship between the mathematical forms in the human mind
called physical theory and physical reality?" Einstein did not believe
in a God who spoke in tongues of flame from the mountaintop in ordinary language, and he could not sustain belief in the anthropomorphic God of the West. There is also no suggestion that he embraced
ontological monism, or the conception of Being featured in Eastern
religious systems like Taoism, Hinduism, and Buddhism. The closest
Einstein apparently came to affirming the existence of the "extra-
personal" in the universe was a "cosmic religious feeling," which he
closely associated with the classical view of scientific epistemology.

The doctrine that Einstein fought to preserve seemed the natural inheritance of physicists until the advent of quantum mechanics. Although the mind that constructs reality might be evolving fictions that are not necessarily true or necessary in social and political life, there was, Einstein felt, a way of knowing purged of deceptions and lies. He was convinced that knowledge of physical reality in physical theory mirrors the pre-existent and immutable realm of physical laws. And, as Einstein consistently made clear, this knowledge mitigates loneliness and inculcates a sense of order and reason in a cosmos that might appear otherwise bereft of meaning and purpose.

What most disturbed Einstein about quantum mechanics was the fact that this physical theory might not in experiment, or even in principle, mirror precisely the structure of physical reality. There is, for reasons we have seen, an inherent uncertainty in measurement of quantum mechanical processes that clearly indicates that mathematical theory does not allow us to predict or know the outcome of events with absolute certainty and precision. Einstein's fear was that if quantum mechanics were a complete theory, it would force us to recognize that this inherent uncertainty applied to all of physics, and, therefore, that the ontological bridge between mathematical theory and physical reality does not exist. This would mean, as Bohr was among the first to realize, that we must profoundly revise the epistemological foundations of modern science.

The lively debate over the epistemological problems presented by quantum physics, reflected in the debate between Einstein and Bohr, came, as the physicist and historian of science Gerald Holton demonstrated, to a grinding halt shortly after World War II.[20] What seems to have occurred, as Holton understands it, is that the position of Einstein became the accepted methodology in physics after World War II, and it has continued to enjoy that status. It is now clear, however, that in the absence of metaphysical assumptions, Einsteinian epistemology can no longer be viewed as valid in scientific terms.

Philosophical Implications of Nonlocality

Although the formalism of quantum physics predicts that correlations between particles over space-like separated regions is possible, it can say nothing about what this strange new relationship between parts (quanta) and whole (cosmos) means outside this formalism. This does not, however, prevent us from considering the implications in philosophical terms, and the implications that may prove the most revolutionary concern the relationship between mind and world. The worldview of classical physics allowed the physicist to assume that communion with the essences of physical reality via mathematical laws and associated theories was possible, but it made no other provisions for the knowing mind. Nature in classical physics was viewed as a vast machine in which forces act between mass points in the abstract background of space and time, and collections of mass points interact as isolated and isolatable systems. The knowing self was separate, discrete, and atomized, and it achieved its knowledge of physical reality from the "outside" of physical systems without disturbing the system under study.

As Henry Stapp put it, "Classical physics not only fails to demand the mental, it fails to even provide a rational place for the mental. And if the mental is introduced ad hoc, then it must remain totally ineffectual, in absolute contradiction to our deepest experience."[21] In addition to providing a view of human beings as cogs in a giant machine linked to other parts of the machine in only the most mundane material terms, classical physics effectively isolated the individual self from any creative aspect of nature. In the classical picture, all the creativity of the cosmos was exhausted in the first instant of creation and what transpires thereafter is utterly preordained.[22]

In our new situation, the status of the knowing mind seems to be very different. All of modern physics contributes to a view of the universe as an unbroken and undissectible dynamic whole. As Melic Capek points out, "There can hardly be a sharper contrast than that between the everlasting atoms of classical physics and the vanishing 'particles' of modern physics."[23] We now know that the classical notion of substance as composed of indestructible building blocks is false and that particles cannot be viewed as separate and discrete. As Stapp put it,

> ...each atom turns out to be nothing but the potentialities in the behavior pattern of others. What we find, therefore, are not elementary space-time realities, but rather a web of relationships in which no part can stand alone; every part derives its meaning and existence only from its place within the whole.[24]

The characteristics of particles and quanta are not isolatable given particle-wave dualism and the incessant exchange of quanta within and between fields. Matter cannot be dissected from the omnipresent sea of energy, nor can we in theory or in fact observe matter from the "outside." As we voyage further into the realm of the unvisualizable in search of the grand unified theory incorporating all forces, the constituents of matter and the basic phenomena involved appear increasingly interconnected, interrelated, and interdependent. As Heisenberg put it, the cosmos "appears as a complicated tissue of events, in which connections of different kinds alternate or overlay or combine and thereby determine the texture of the whole."[25]

This means that a purely reductionist approach to understanding physical reality, which was the goal of classical physics, is no longer appropriate. For reasons other than those provided by quantum indeterminacy, this should have been obvious when Kurt Gödel published his incompleteness theorem in 1930. Even if we develop a "Theory of Everything," and even if that theory coordinates within its mathematical framework an explanation of the phenomena of life and/or consciousness, this theory could not in principle claim to be the final or complete description.

In our view, the general principle of complementarity, which applies to all scales and times, is a complement to the incompleteness theorem. Within the framework of this general principle, no mathematical theory can be complete because reality-in-itself can never be finally disclosed or defined. No collection of parts will ever constitute the whole, and the properties of parts can only be understood in terms of embedded relations within the whole.

Parts and Wholes in Physical Reality

As the philosopher of science Earl Harris noted in thinking about the special character of wholeness in modern physics, a "whole is always and necessarily a unity of and in difference."[26] A unity without internal content is a blank or empty set and is not recognizable as a whole. A collection of merely externally related parts does not constitute a whole because the parts will not be "mutually adaptive and complementary to one another." Wholeness requires a complementary relationship between unity and difference, and is governed by a principle of organization determining the interrelationship between parts. This organizing principle must be universal to a genuine whole and implicit in all parts that constitute the whole, even though the whole is exemplified only in its parts. This principle of order, according to Harris, "is nothing real in and of itself. It is the way the parts are organized, and not another constituent additional to those that constitute the totality."[27]

In a genuine whole, the relationships between the constituent parts must be "internal or immanent" in the parts, as opposed to a more spurious whole in which parts appear to disclose wholeness due to relationships that are "external" to the parts.[28] The classical view of the whole as consisting of the sum of externally related parts is an example of a spurious whole. Parts constitute a genuine whole when the universal principle of order is in some sense "inside" the parts, and thereby adjusts each to all so that they interlock and become mutually complementary. The terms and relations between parts in such a whole have in common the immanent principle of order that is the "structure of the system to which they belong."[29] This not only describes the character of the whole revealed in both relativity theory and quantum mechanics; it is also consistent with the manner in which we have begun to understand the relation between parts and whole in modern biology.

On the most primary level, the cosmos is a dynamic sea of energy manifesting itself in entangled quanta and seamlessly interconnected events. This is true for the contentless content of the singularity at the origins of the cosmos and throughout all subsequent stages of cosmic evolution. This dynamic progress eventually unfolds into a complementary relationship on the most complex of levels between organic molecules, which display biological regularities, and inorganic molecules, which do not. Additional part-whole complementarities emerged as biological life became more complex and regulation of global temperature and relative abundance of atmospheric gases by the whole was required to sustain this life.

Modern physics also reveals a complementary relationship between the differences between parts that constitute content and the universal ordering principle that is immanent in each of the parts. Although the whole cannot be finally disclosed in the analysis of the parts, the study of the differences between parts provides insights into the dynamic structure of the whole present in the parts. The part can never, however, be finally isolated from the web of relationships that discloses the interconnections with the whole, and any attempt to do so results, as we have seen, in ambiguity.

Much of the ambiguity in attempts to explain the character of wholes in both physics and biology derives from the assumption that order exists "between" parts. But order in the complementary relationship between difference and sameness in any physical event is never external to that event—the connections are immanent in the event. That nonlocality would be added to this picture of the dynamic whole with Bell's theorem and the experiments testing that theorem is not really surprising. The relationship between part, as quantum event apparent in observation or measurement, and the undissectable whole, revealed but not described by the instantaneous correlations between measurements in space-like separated regions, is another example of part-whole complementarity in modern physics.

Each of the systems we attempt to isolate in the study of nature is in some sense a whole because each system represents the whole in the activity of being the part. But no single system, with the exception of the entire universe, can fully realize the cosmic order of the totality due to the partial and subordinate character of differentiated systems. No part can sustain itself in its own right because difference is only one complementary aspect of its being; the other aspect requires participation in the sameness of the cosmic order. All differentiated systems in nature require, in theory and in fact, supplementation by other systems,[30] and this now appears to be as true in biology as it is in physics.

If we accept this view of higher, or more progressive, order in the universe, then each stage in the emergence of complementary relations may allow for more complex realizations of the totality than that which has gone before.[31] In other words, progressive order in this sense is a process in which every phase is a summation and transformation of the previous phase. The emergent order apparent in a new phase results from the incorporation of previous phases, or transformations, through which it has developed and as a result of which it assumes its own form or structure. This means that process and end could be wed in a difference/sameness complementarity with the organizing principle of the whole immanent throughout the proliferation of the constituent parts.

This principle seems immanent from the beginning and discloses its potential, according to Harris, in the "entire scale of systematically (internally) related phases."[32] From this perspective, the best scientific description of the cosmos that we can ever achieve will only approximate the realization of the totality apparent to date.[33] The ultimate organizing principle resides, concluded Harris, "in the outcome of the process and not its genetic origin." Hence, reductionism is "ruled out" and will remain so even if quantum physics provides a satisfactory account of the origins of the cosmos.[34]

The Conscious Universe

In these terms, consciousness, or mind, can be properly defined as a phase in the process of the evolution of the cosmos implied in presupposing all other stages.[35] If consciousness manifests or emerges in the later stages and has been progressively unfolding from the beginning stages, we can logically conclude, as opposed to scientifically prove, that the universe is conscious. In the grand interplay of quanta and field in whatever stage of complexity, including the very activities of our brain, there is literally "no thing" that can be presumed isolated or discrete. From this perspective, consciousness could be an emergent phenomenon that folds within itself progressive stages of order embedded in part-whole complementarities throughout the history of the universe.

We do not mean to suggest that this consciousness is in any sense anthropomorphic. Our current understanding of nature neither supports nor refutes any conceptions of design, meaning, purpose, intent, or plan associated with any mytho-religious or cultural heritage. As Heisenberg put it, such words are "taken from the realm of human experience" and are "metaphors at best." What we mean by "conscious universe" is, however, consistent with the totality of scientific facts and is anthropocentric only to the extent that it answers a very basic human need. The need is to feel that a profound awareness of unity with the whole is commensurate with our scientific world-view and cannot be discounted or undermined with an appeal to scientific knowledge.

This knowledge indicates that on the most fundamental level the universe evinces an undivided wholeness, and this wholeness in modern physical theory does appear to be associated with a principle of cosmic order. If this principle were not a property of the whole that exists within the parts, it seems reasonable to conclude that there would be no order or no higher-level organization of matter that allows for complexity. Because the whole, or reality-in-itself, transcends space-time and exists or manifests within all parts or quanta in space-time, the principle of order seems to operate in self-reflective fashion. If the whole were not self-reflectively aware of itself as reality-in-itself, the order that is a precondition for all being would not, in our view, exist. Since human consciousness in its most narrow formulation can be defined as self-reflective awareness founded on a sense of internal consistency or order, we can infer, but not prove, that the universe is, in these terms, conscious.

Consciousness and the Single Significant Whole

The apprehension of the single significant whole as it is disclosed in physical theory and experiment may indicate that we have entered another, more advanced "stage" in the evolution of consciousness. What this theology of mind, or consciousness, "assumes" is the progressive realization of the totality of the organizing principle. This view becomes particularly compelling when we consider that complementary constructs appear to be as foundational to our conscious constructions of reality in mathematical languages as they are to the unfolding of progressive stages of complexity in physical reality.

The radical thesis here is that human consciousness may fold within itself the fundamental logical principle of the conscious universe. Perhaps this prior condition in human consciousness allowed us to construct a view of this universe in physical theory that describes the unfolding of the cosmic order at previous stages in the life of the cosmos. In this view, the human mind is not a "disease of matter" in an "alien" universe that happened to discover that the activities of this universe can be better understood in terms of mathematical relations in physical theory. It is an emergent phenomenon in a seamlessly interconnected whole that contains within itself the fundamental logical principle embedded in the activities of this whole.

In our view, one of the most dramatic indications that this might be the case is the three types of nonlocalities discussed in the last chapter. Although we can confirm the existence of type I (spatial) and type II (temporal), these complementary constructs can only infer the existence of the undivided wholeness represented by type III. The two nonlocalities that can be confirmed by experiment may bring us to the horizon of knowledge where we confront the existence of the undivided whole, but we cannot cross that horizon in terms of the content of consciousness. The fact that we cannot disclose this undivided wholeness in our conscious constructions of this reality as "parts" does not mean that science invalidates the prospect that we can apprehend this wholeness on a level that is prior to the conscious constructs; it merely means that science qua science cannot fully disclose or describe the whole.

This situation allows a rational place for the mental and thereby demolishes the notion that the realm of human thought and feeling is merely ad hoc. Life and cognition in these terms can be viewed as "grounded" in the single significant whole, but this is not a conclusion that can be "proven" in strictly scientific terms for all the reasons discussed. This single significant whole must be represented in the conscious content as parts, and cannot, therefore, be a direct object of scientific inquiry or knowledge. Thus, any direct experience we have of this whole is necessarily in the background of consciousness and must be devoid of conscious content.

Herein lies the paradox. Scientific knowledge conditions us into an awareness of the possible existence of the single significant whole and yet cannot fully affirm or "prove" its existence. This would also seem to argue, however, that all those who apprehend the single significant whole or experience cosmic religious feeling, with or without the awareness of the existence of the principle of cosmic order, are engaged in similar acts of communion with the whole. Any translation into conscious content of that experience in scientific or religious thought, however, invokes reductionism where it cannot be applied. Put differently, all knowledge in the conscious content is a differentiated system that cannot by definition articulate the universal principle of order.

A New Dialog Between Science and Religion

For those so inclined, there is a new basis for dialog between science and religion. If this dialog were as open and honest as it could and should be, we might begin to discover a spiritual pattern that could function as the basis for a global human ethos. Central to this vision would be a cosmos rippling with tension evolving out of itself endless examples of the awe and wonder of its seamlessly interconnected life. Central to the cultivation and practice of the spiritual pattern of the community would be a profound acceptance of the astonishing fact of our being.

Religious thinkers can enter this dialog with the recognition that because science can no longer legislate over the ontological question in classical terms, the knowledge we call science cannot in principle be used to dismiss or challenge belief in spiritual reality. However, if these thinkers elect to challenge the truths of science within its own domain, they must either withdraw from the dialog or engage science on its own terms. Applying metaphysics where there is no metaphysics, or attempting to rewrite or rework scientific truths and/or facts in the effort to prove metaphysical assumptions, merely displays a profound misunderstanding of science and an apparent unwillingness to recognize its successes. However, it is also true that the study of science can serve to reinforce belief in profoundly religious truths while not claiming to legislate over the ultimate character of these truths.

If the dross of anthropomorphism can be eliminated in a renewed dialog between the two kinds of truth, the era in which we were obliged to conceive of these truths as two utterly disparate truths and, therefore, as providing no truth at all, could be over. In our new situation, science in no way argues against the existence of God, or Being, and it can profoundly augment the sense of the cosmos as a single significant whole. That the ultimate no longer appears to be clothed in the arbitrarily derived terms of our previous understanding may simply mean that the mystery that evades all human understanding remains. The study of physical reality will, in our view, only take us perpetually closer to that horizon of knowledge where the sum of parts is not and cannot be the whole.

Wolfgang Pauli, who thought long and hard about the ethical good that could be occasioned by a renewed dialog between science and religion, made the following optimistic forecast:

> Contrary to the strict division of the activity of the human spirit into separate departments—a division prevailing since the nineteenth century—I consider the ambition of overcoming opposites, including also a synthesis embracing both rational understanding and the mystical experi-

ence of unity, to be the mythos, spoken and unspoken, of our present day and age.[36]

This is a project that will demand a strong sense of intellectual community, a large capacity for spiritual awareness, a profound commitment to the proposition that knowledge coordinates experience in the interest of survival, and an unwavering belief that we are free to elect the best means of our survival. The essential truth revealed by science that the religious imagination should now begin to explore with the intent of enhancing its ethical dimensions was described by Schrödinger as follows:

> Hence this life of yours which you are living is not merely a piece of the entire existence, but is, in a certain sense, the whole; only this whole is not so constituted that it can be surveyed in one single glance.[37]

Virtually all major religious traditions have at some point featured this understanding in their mystical traditions, and the history of religious thought reveals a progression toward the conception of spiritual reality as a unified essence in which the self is manifested, or mirrored, in intimate connection with the whole. Although some have derived this profound sense of unity based only on a scientific worldview, or in the absence of metaphysics, most people, as Schrödinger notes, require something more:

> The scientific picture of the real world around me is very deficient. It gives me a lot of factual information, puts all our experience in a magnificently consistent order, but it is ghastly silent about all and sundry that is really dear to our heart, that really matters to us.[38]

It is time, we suggest, for the religious imagination and religious experience to engage the complementary truths of science in filling that silence with meaning. The question of belief in ontology remains what it has always been, a question, and the physical universe on the most basic level remains what it has always been, a riddle. The ultimate answer to the question and the ultimate meaning of the riddle are, and probably always will be, a matter of personal choice and conviction. That choice is now, however, freed from the constraints imposed on it by the scientific description of nature since the eighteenth century. The universe more closely resembles a great thought than a great machine, and human consciousness may be one of its grander manifestations.

Appendix
Horizons of Knowledge in
Cosmological Models

As discussed earlier, in any cosmological model of the universe, including big bang models, we encounter horizons of knowledge that arise because a particular model has been adopted. For big bang models, the most obvious horizon has to do with the fact that the universe becomes opaque to its own radiation for sufficiently early times. The best way to describe the situation is in terms of the redshift. We can express the age of the universe as a function of the fractional shift in wavelength, or redshift z; z is defined as the ratio between the change of the wavelength, $\lambda_o - \lambda_e$, over the emitted wavelength λ_e, and λ_o is the observed wavelength seen by a moving observer. One hundred thousand years after the beginning, when the universe was only 0.1 percent of 1 percent of its current age, corresponds to z of 1000.

The most distant quasars, seen at redshifts of 4 to 5, emitted their observable light when the universe was about ten percent of its current age. Although radiation can in principle tell us much more about the early universe than matter can, the opaqueness of the universe prior to $z = 1000$ does not allow us to trace or confirm the origin of the universe in the big bang cosmology using photons. However, photons provide virtually all information about both the large-scale structure of the universe and what we know of its evolution. It is, therefore, at $z = 1000$ that we encounter the first horizon of knowledge about our universe in any big bang theoretical model. If our only access to reliable information about the earlier universe must be based on the observation of photons, that horizon is, in principle, impregnable.[1]

Another possibility would be to use neutrinos as a probe for the early universe. Some particle physicists have postulated that neutrinos possess a rest mass with a value that, although small, is not exactly zero and, therefore, that all the neutrinos in the universe together could provide enough "missing mass" to close the universe. Recent laboratory evidence suggests that neutrinos possess a small mass. If primordial neutrinos emitted a few seconds after the beginning of the universe, or at a redshift $z \sim 10^9$, were ever observed, that would be exciting. Although it is anticipated that their numbers would be much greater than those produced by stellar events such as supernova explosions, their energies would be billions of times less. But even if the problem of detecting primordial neutrinos could be overcome, this would not yield sufficient information about the very early universe to answer fundamental questions. The problem is that the very high redshift $z \sim 10^9$, at which primordial neutrinos would have been emitted, is not sufficiently large to allow us to probe time scales near the big bang itself or near the hypothetical inflationary period of 10^{-35}. This means that the observational horizon of knowledge at $z \sim 10^9$ is the ultimate horizon from which we can access direct information about the universe.[2]

The other major source of information about the early life of the universe is experiments in particle accelerators. Although studies of ordinary matter in high-energy physics can yield a lot of detailed information about the hypothesis of element formation in the early universe,[3] the actual situation is not so rosy. The uncertainties in the abundances of the primordial elements, like deuterium, regular helium, and Li, are large in the big bang models. In addition, the details of big bang models are least sensitive to the abundance of regular helium, which is by far the most abundant of primordial elements formed from hydrogen (see Fig. 29). Current results, taken at face value, imply a mean density of baryons about 3×10^{-31} gr/cm^3, about two orders of magnitude less than the critical density required to close the universe. What is most significant here is that if the only matter in the universe is luminous matter made of baryons, the results imply an open universe. However, even if we could somehow identify in the future what part of these elements was primordial, or synthesized in the first three minutes, we could still not go further back in redshift than $z = 10^8$.

Of all these horizons of knowledge at $z = 10^9$, 10^8, and 1000, it is the last that is likely to remain the only one we can explore for the foreseeable future. The neutrino horizon at 10^9 is not going to be quantitatively different from the photon horizon. Because both photons and neutrinos at early periods were in thermal equilibrium with matter, both primordial neutrinos and photons follow a black body distribution because they were in equilibrium with themselves and everything else early on. It seems clear that whatever problems of interpretation we are facing today with regard to the background photons will not go away even if we manage to observe primordial neutrinos. We should also not anticipate that high energy physics will probe the element horizon before the first three minutes. The problem here is that we cannot simulate the complexity of nuclear reactions applicable at that time in the early universe due to the inherent uncertainties of the reactions themselves.

The last observational problem in big bang cosmology we want to mention involves the measurements of redshifts of the light spectra of distant galaxies. In the effort to test the geometry of the universe, one studies the Hubble diagram, or a diagram of magnitude versus redshift using "standard candles"—distant galaxies that supposedly have a luminosity that does not vary with time. One could determine the curvature effects of the universe in the Hubble diagram. Although quasars can be seen as far away as $z = 4$–5, they are notoriously unreliable as standard candles because of the tremendous range of luminosities they present for any redshift.

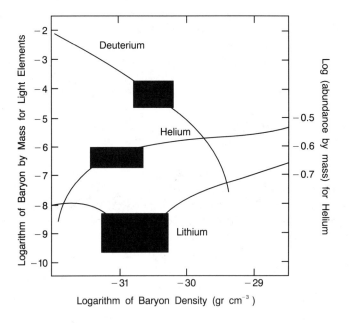

FIGURE 29. Abundances of primordial light elements. Taken at face value, these results imply an open universe with density of matter in the range 10^{-31}–10^{-30} gr cm^{-3}.

The traditional approach has been to search for other standard candles. It would seem that the best choice would be ordinary galaxies, like the Milky Way. Most ordinary galaxies cannot be observed from the ground beyond a redshift of about unity and are now observable through the Hubble Space Telescope. In the past, astronomers relied on bright radio galaxies as standard candles, which are much further away than ordinary galaxies. One of the most distant galaxies discovered so far has a redshift of 3.8 at a corresponding distance of 15 billion light years.[4] Unfortunately, radio galaxies have turned out not to be "standard" after all[5,6]; their light changes significantly with distance and/or distant radio galaxies are different from nearby ones. It also remains to be seen whether distant ordinary galaxies will turn out to be "ordinary" after all. It may turn out that all galaxies look sufficiently different when the universe was ten percent of its age compared to today. Redshifts in the range 5 to 10 are particularly important to study because astronomers suspect that it is in this range where we would detect light from galaxies that began to form more than 15 billion years ago.

Also, one requires redshifts of this magnitude to distinguish which big bang model is applicable when one studies the Hubble diagram (Fig. 30) because the observational uncertainties are so large that one cannot distinguish between competing models for smaller redshifts. Even if we assume that the quest for "standard candles" is successful, we still encounter a difficulty that may turn out to be insurmountable—images of ordinary galaxies begin to merge at smaller redshifts.[7] This can be easily shown if one examines how the apparent size of a finite-size object varies with distance in the big bang cosmology.

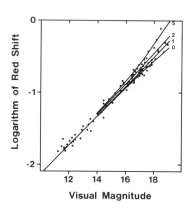

FIGURE 30. The Hubble diagram and the various associated geometries for distant galaxies. The curves are labeled according to a parameter: 0–1/2 (open); 1/2 (Einstein-de Sitter); >1/2 (closed). (Data from M. Rowan-Robinson, *The Cosmological Distance Ladder*, New York: W.H. Freeman, 1985.)

It is known that the apparent size of an object of some "standard" length, such as a galaxy 60,000 light years across, decreases as the distance increases. General relativity predicts that above a certain distance, which in most big bang models corresponds to a redshift of about one, the apparent size will continue to increase as the distance increases. This happens because of curvature effects. At some redshifts, which turn out to be high but still less than ~50, the apparent size of a distant galaxy tens of thousands of light years across would be a few arc-seconds. This apparent size is comparable to the distance between this galaxy and other neighboring galaxies, also a few arc-seconds. This conclusion is based on extrapolating the observed distribution of faint galaxies in the sky to fainter and fainter magnitudes. When we do so, the exact value of the redshift at one magnitude is subject to uncertainties about the distant and currently unobserved ordinary galaxies. If this conclusion is borne out by future space-borne observations, it will be observationally impossible to obtain an accurate spectrum of any galaxy beyond redshifts at less than ~50. And this would lead to the even more dramatic conclusion that is impossible to test which of the competing big bang models is applicable.

The point, which will undoubtedly disturb many cosmologists, is that the geometry of the universe and the appropriate applicable model will be indeterminate. Since the "galaxy image horizon" is fairly close to us, this suggests that the experimental methods in use since the time of Hubble, which have led to the hypothesis of the expanding universe, may not allow this hypothesis to be fully verified or proven. Spectra of different sources in the sky would themselves be blended together as one looks at fainter and fainter sources and the images begin to merge. Eventually, the background from different galaxies would dominate the spectrum from a single distant galaxy, and reliable spectra could not be obtained to test the geometry of the universe.

Hypothesis of Large-Scale Complementarities

In our view, the reason the universe is observed to be so close to the flat, or Euclidean, model is that this model represents the division between the complementary constructs of open and closed cosmologies. Rather than attempting to understand the observed flatness in terms of a single construct, which requires that we introduce ad-hoc constructs like dark matter and the cosmological constant, the complementary constructs of open and closed cosmologies resolve apparent ambiguities and result in a more coherent picture. As our observations carry us to deeper regions of space, the inherent uncertainties of specific big bang models, such as when galaxies were formed or how distant galaxies differ from nearby galaxies, will not allow us to unequivocally prove the correctness of one of the cosmological models. Put differently, the observational errors which arise from the experimental situation coupled with the inherent theoretical uncertainties of the cosmological models themselves will not allow unique tests to be performed that confirm the validity of one of the models. More simply put, there will always be alternative ways to interpret the observational results in cosmology.

If, for example, we find a peculiar property of galaxies at, say, a redshift $z = 3$, we would need to go further back in time to higher redshifts to see whether this property persists to earlier times. But going further back to earlier times results in greater observational uncertainties since light from more distant sources will be blended together. If we make additional theoretical assumptions in an effort to get around this problem, they will not work for a now obvious reason—the assumptions would not be directly testable. Here we confront the essential problem in our dealings with all observations of cosmological import. Experience has shown that as our observations get better, we push our knowledge further back to times where the predictions of particular theories are in conflict and where it is impossible to make observations that could determine which theory is correct.

If the flatness, horizon, and isotropy problems cannot be resolved by either theory or observation, the obvious question is why measurement leads to logically disparate constructs describing the universe. This is a situation in which we seemingly cannot, in principle, determine whether the universe is open or closed, whether the big bang or the steady-state model is correct, whether the redshifts are cosmological or not, or whether the constants of nature are "constant" or, as Dirac[8] was the first to suggest, vary over time. Also, H. Alfven and others[9] have suggested that the universe might not have started from a big bang and that gravity does not dominate the large-scale structure of the universe. Yet there is also the alternate hypothesis that the electromagnetic effects in the cosmic plasma determine the structure of the universe, including the observed superclusters. The following are possible profound complementarities that, we believe, may be disclosed in our future studies of the large-scale structure of the universe:

- open big bang model/closed big bang model
- big bang cosmology/steady-state cosmology
- cosmological redshifts/discordant redshifts

- constants of nature are variable/"constants" are varying
- big bang cosmology/plasma cosmology

In our view, what we confront here is not a "practical limit" to observation. Rather, we confront conditions for observation comparable to those that result in wave-particle dualism. Here, as in laboratory experiments in quantum physics, we confront a situation in which the observer and his observing apparatus must be taken into account. If we view open and closed models in big bang cosmology in these terms, it seems clear that these seemingly profound oppositions are, in actuality, complementary constructs. In the discussion that follows, we will attempt to illustrate that the arguments that support the hypothesis that open and closed models are complementary constructs can also be applied to the other candidates listed above.

Any big bang model of the universe, open or closed, results in the observational horizon that appears a few hundred thousand years after origins during the period when the universe was opaque to its own radiation. Any particular model is ambiguous for earlier times and any appeals to observation cannot resolve the inherent ambiguities. Although it also seems clear that any single and specific theoretical model requires the existence of this observational horizon, this horizon *is not necessarily inherent or of the same type in different theoretical cosmological models*. If one, for example, adopts the steady state cosmological theory of Alfven's plasma cosmological theory, which assumes no beginning, and, therefore, no relic background radiation, one does not confront the big bang problem of optical thickness at $z = 1000$. But if the redshifts are not cosmological, as Halton Arp, Geoffrey Burbidge, Fred Hoyle, and others have argued, this non-cosmological assumption would eliminate the $z = 1000$ optical thickness observation problem encountered in the big bang models.

Halton Arp[10] has found numerous examples where an apparent association between a galaxy of a small redshift and a quasar of a different and generally large (or "discordant") redshift occurs. These associations may be chance associations of nearby galaxies and more distant quasars that happen to be along the same lines of sight, even though a number of gaseous bridges have been found that would indicate spatial associations. Physicist Emil Wolf[11] and others[12] have proposed a plausible alternative physical mechanism to explain galaxy redshifts. These alternatives to the big bang are not in vogue due to the dominance of the big bang paradigm. But one has to remember that the big bang theory is just a theory, as opposed to a self-evident truth, and is, therefore, subject to observational scrutiny. Yet these other models also present their own horizons of observability, which are not the same as those encountered in the big bang model.

If, as noted earlier, the conditions for observation that result in the open/closed models are analogous to those that result in wave-particle dualism, perhaps the two models should be viewed as complementary aspects of the complete big bang theory of the universe. It is already clear in cosmology that observations yield results that can be used to support each of these logically antithetical models. When one attempts to add up all the luminous matter in the universe, one obtains insufficient mass to "close" the universe. Hence the universe appears to be open and, as recent Hubble Space Telescope observations indicate, could be accelerating. On the other hand, when one attempts to study the behavior of the Hubble Law for distant galaxies, one finds a tendency toward a closed model. Although the errors in this second study are large, we are predicting that future improved observations will not resolve the existing ambiguity. In fact, what distinguishes our approach from other approaches, which attempt to explain the observations by forcing theoretical models, is that we are predicting that the ambiguities will not go away. They will remain as long as one insists on attempting to explain all observations within a single model.

Quantum Resolution of Cosmological Measurement Problems

If we can divest ourselves of the distorting lenses of classical assumptions, the case that the universe is a quantum system at all scales and times is easily made. If this conclusion is as factual as we understand it to be, we can address the problems encountered by the big bang cosmology in terms of first principles, or without appealing to an ad hoc model like inflation. There is, first of all, general agreement among virtually all cosmologists and quantum physicists that all quanta were entangled early on in the history of the universe. If that is the case, we must also conclude that this entanglement should remain a frozen-in property of the macrocosm.[13] Because quantum entanglement in the experiments testing Bell's theorem reveals an underlying wholeness that remains a property of the entire system even at macroscopic distances, the seemingly inescapable conclusion is that the underlying wholeness associated with quantum entanglement in the early universe remains a property of the universe at all times and all scales. If this is the case, it would seem to imply that any subsequent interaction with quanta at any stage in the life of the universe could reveal emergent complementarities.

In attempting to prove the "correctness" of one of the disparate constructs of open or closed universe, we have to observe increasingly

distant sources of light. Bigger ground telescopes and space telescopes are being built to collect the photons from more distant, faint
sources. One cannot, however, ultimately avoid the problems arising
from the complementary nature of light. If the expected photons are
few, one confronts the following dilemmas: (1) If we disperse the light
to obtain a spectrum, the exact location of the faint source of light at
this cosmological distance would be indeterminate; or (2) if we elect to
photographically record the position of the faint source, the spectrum
at this cosmological distance would be unknown.

One could argue that this is not a problem when two observations
are made—one in which the spectrum is obtained and another in
which the position is obtained. This does not, however, eliminate the
observational uncertainties. Due to the faintness of the presumed
source of light and its apparent nearness to other faint sources of
light, one could not be sure that the photons in the first experiment
emanated from the same exact location in the sky as the photons in
the second experiment. Dispersing more of the light to obtain a better
spectrum would exacerbate the problem, as the uncertainty principle
implies, because the direction of origin of the observed photons in the
sky would become more uncertain. If the quantum nature of light has
to be taken into account in observations of cosmological import, and if
the entanglement of quanta in the early universe remains a frozen-in
property at all scales and times, one would not be surprised to encounter horizons of knowledge.

If this is the case, it is not unreasonable to conclude that the flatness, horizon, and isotropy problems of the big bang models could be
the direct result of limits of observation in the quantum universe
which call for the application of complementary theoretical constructs. Perhaps what makes the observations problematic is that
cosmologists have treated them as preconditions for the confirmations
of single theoretical models that compete with one another for complete domination of the situation. These "problems" and others that
future observations in cosmology are likely to reveal could, therefore,
be eliminated with the simple realization that quantum entanglement remains a frozen-in property of the cosmos, even at "macroscopic" scales and for all times.

One can then conclude that these problems arise as a result of
making observational choices in observing a quantum system, and,
therefore, are not really problems. Perhaps what we are actually confronting here is merely additional demonstrations of the underlying
wholeness implicit in this system. In our view, the reason the universe is so close to the flat, Euclidean model is that the model represents the division between the complementary constructs of open and
closed cosmologies. As our observations carry us to fainter sources,
which in the big bang model correlates with more distant regions of

space, the observational uncertainties may be testifying to our inability to apply a single model where another logically disparate but complementary model is also required for a complete view of the situation. However one feels about this argument—and some feelings are likely to be strongly negative—the observational evidence in cosmology suggests that there will "always" be alternative ways to interpret the results. Experience has shown that as our observations get better, we push our knowledge further back to times where theory cannot provide unequivocal predictions that can become the basis for further tests.

One also faces the problem of initial conditions for the universe. Attempts by Stephen Hawking and others to circumvent this problem by appealing to the "no boundary" proposal ontologize the wave function of the universe and are not, for all the reasons discussed in the chapter on quantum ontologies, subject to experimental verification. At this point, it seems reasonable to conclude that observational horizons of knowledge have left us in a situation in which we seemingly cannot, in principle, determine whether the universe is open or closed. The majority view is probably that we have merely confronted a practical limit to observation that will be eliminated by future improvements in observational technologies and techniques. Our alternative view is that what we are actually confronting are conditions for observation comparable to those that invoke wave-particle dualism in quantum physics.

If this is the case, the reason that we cannot decide whether the open or closed model is correct is rather obvious—the logically disparate constructs are complementary. Although one precludes the other in application to a particular situation, both are required for the complete description. In this view, what is prior to any particular measurement of this reality is the same thing that is prior to measurement in all quantum mechanical experiments—the undivided wholeness of reality-in-itself. Yet the wholeness in the universe we have in mind here is not an a priori philosophical assumption but an emergent property of a quantum universe that reveals itself under experimental conditions that clearly indicate that the observer cannot be independent of the observing process at any level.

Notes

Introduction

[1]W. Tittel, J. Brendel, H. Zbinden, and N. Gisin, "Violation of Bell Inequalities by Photons More than 10 km Apart," *Physical Review Letters*, 26 October 1998, *81*, no. 17, pp. 3563–6.
[2]N. David Mermin, "Extreme Quantum Entanglement in a Superposition of Macroscopically Distinct States," *Physical Review Letters*, October 8, 1990, *65*, no. 15, pp. 1838–40.
[3]Frederick Suppe, *The Structure of Scientific Theories* (Chicago: The University of Illinois Press, 1977), p. 126.
[4]Ibid, pp. 694–5.

Chapter 1

[1]Felix Browder, *Mathematical Developments Arising from Hilbert Problems,* Proceedings of Symposia in Pure Mathematics, vol. 28 (Providence, R.I.: American Mathematical Society, 1974).
[2]Albert Einstein, "How I Created the Relativity Theory," lecture in Kyoto, Japan, 14 December 1922, in Abraham Pais, *Subtle is the Lord* (New York: Oxford University Press, 1987), p. 131.
[3]Henri Poincare, *Science and Hypothesis*, trans. W.J.G. (New York: Dover, 1952), p. 90.
[4]Ibid, pp. 168, 176.
[5]Werner Heisenberg, quoted in James B. Conant, *Modern Science and Modern Man* (New York: Columbia University Press, 1953), p. 40.
[6]Ernest Rutherford, quoted in Ruth Moore, *Niels Bohr: The Man, His Science and the World They Changed* (New York: Knopf, 1966), p. 40.
[7]Werner Heisenberg, quoted in Conant, *Modern Science and Modern Man*, p. 40.
[8]Max Jammer, *The Conceptual Development of Quantum Mechanics* (New York: McGraw-Hill, 1966), p. 271.
[9]Werner Heisenberg, quoted in Conant, *Modern Science and Modern Man*, p. 271.
[10]Robert Oppenheimer, quoted in Ibid.
[11]Clifford A. Hooker, "The Nature of Quantum Mechanical Reality," in

Paradigms and Paradoxes, ed. Robert Colodny (Pittsburgh: University of Pittsburgh Press, 1972), p. 132.

Chapter 2

[1]Eugene P. Wigner, "The Problem of Measurement," in *Quantum Theory and Measurement*, ed. John A. Wheeler and Wojciech H. Zurek (Princeton, N.J.: Princeton University Press, 1983), p. 327.
[2]Ibid, p. 327.
[3]Olivier C. de Beauregard, private communication (1988).
[4]Richard Feynman, *The Character of Physical Law* (Cambridge, Mass.: MIT Press, 1967), p. 130.
[5]John A. Wheeler, "Beyond the Black Hole," in *Some Strangeness in the Proportion*, ed. Harry Woolf (London: Addison-Wesley, 1980), p. 354.
[6]See Abner Shimony, "The Reality of the Quantum World," *Scientific American*, January, 1988, p. 46.
[7]Richard P. Feynman, *QED: The Strange Theory of Light and Matter* (Princeton, N.J.: Princeton University Press, 1985), p. 7.
[8]Ibid, p. 25.
[9]See Paul C.W. Davies, *Quantum Mechanics* (London: Routledge & Kegan Paul, 1984).
[10]Richard P. Feynman, *The Character of Physical Law*, p. 80ff.
[11]Abner Shimony, "The Reality of the Quantum World," p. 48.
[12]Steven Weinberg, quoted in Heinz Pagels, *The Cosmic Code* (New York: Bantam Books, 1983), p. 239.

Chapter 3

[1]Abraham Pais, *Subtle is the Lord* (New York: Oxford University Press, 1982).
[2]A. Einstein, B. Podolsky, and N. Rosen, "Can Quantum-Mechanical Description of Physical Reality be Considered Complete?" *Physical Review*, 1935, *47*, p. 777. Paper was reprinted in *Physical Reality*, ed. S. Toulmin (New York: Harper and Row, 1970).
[3]Albert Einstein, Ibid.
[4]Nick Herbert, *Quantum Reality: Beyond the New Physics, An Excursion into Metaphysics and the Meaning of Reality* (Garden City, N.Y.: Anchor Press, 1987), p. 216ff.
[5]Bernard d'Espagnat, *Physical Review Letters*, 1981, *49*, p. 1804.
[6]See A. Aspect, J. Dalibard, and G. Roger, *Physical Review Letters*, 1981, *47*, p. 460.
[7]See Henry P. Stapp, "Quantum Physics and the Physicist's View of Na-

ture: Philosophical Implications of Bell's Theorem," in *The World View of Contemporary Physics*, ed. Richard E. Kitchener (Albany, N.Y.: S.U.N.Y. Press, 1988), p. 40.

[8]D.M. Greenberger, M.A. Home, and A. Zeilinger, "Going Beyond Bell's Theorem," in *Bell's Theorem, Quantum Theory and Conceptions of the Universe*, ed. M. Kafatos (Dordrecht, Holland: Kluwer Academic Publishers, 1989), pp. 69–72.

[9]See Bernard d'Espagnat, *In Search of Reality* (New York: Springer-Verlag, 1981), pp. 43–8.

[10]N. David Mermin, "Extreme Quantum Entanglement in a Superposition of Macroscopically Distinct States" *Physical Review Letters*, October 3, 1990, *65*, no. 15, pp. 1838–40.

Chapter 4

[1]Niels Bohr, *Atomic Theory and the Description of Nature* (Cambridge, England: Cambridge University Press, 1961), pp. 4, 34.

[2]See Clifford A. Hooker, "The Nature of Quantum Mechanical Reality," in *Paradigms and Paradoxes*, pp. 161–2. Also see Niels Bohr, *Atomic Physics and Human Knowledge* (New York: John Wiley and Sons, 1958), pp. 26, 34, 72, 88ff, and *Atomic Theory and the Description of Nature*, (Cambridge, England: Cambridge University Press, 1961) pp. 5, 8, 16ff, 53, 94.

[3]Niels Bohr, "Causality and Complementarity," *Philosophy of Science*, 1960, *4*, pp. 293–4.

[4]Ibid.

[5]Clifford A. Hooker, "The Nature of Quantum Mechanical Reality," p. 137.

[6]See Abraham Pais, *Subtle Is the Lord* (New York: Oxford University Press, 1982), p. 456.

[7]Leon Rosenfeld, "Niels Bohr's Contributions to Epistemology," *Physics Today*, April 29, 1961, *190*, p. 50.

[8]Niels Bohr, *Atomic Theory and the Description of Nature*, pp. 54–5.

[9]Niels Bohr, "Discussions with Einstein on Epistemological Issues," in Henry Folse, *The Philosophy of Niels Bohr: The Framework of Complementarity* (Amsterdam: North Holland Physics Publishing, 1985), pp. 237–8.

[10]Niels Bohr, *Atomic Physics and Human Knowledge*, pp. 64, 73. Also see Clifford A. Hooker's detailed and excellent discussion of these points in "The Nature of Quantum Mechanical Reality," pp. 57–302.

[11]Clifford A. Hooker, "The Nature of Quantum Mechanical Reality," p. 155.

[12]Niels Bohr, *Atomic Physics and Human Knowledge*, p. 74.

[13]Niels Bohr, *Atomic Theory and the Description of Nature*, pp. 56–7.

[14]Niels Bohr, *Atomic Physics and Human Knowledge*, p. 74.

[15]Niels Bohr, "Physical Science and Man's Position," *Philosophy Today* (1957), p. 67.

[16]Leon Rosenfeld, "Foundations of Quantum Theory and Complementarity," *Nature*, April 29, 1961, *190*, p. 385.

[17]Niels Bohr, *Atomic Physics and Human Knowledge*, p. 79.

[18]Niels Bohr, quoted in A. Peterson, "The Philosophy of Niels Bohr," *Bulletin of the Atomic Scientists*, September 1963, p. 12.

[19]Niels Bohr, *Atomic Theory and the Description of Nature*, p. 49.

[20]Albert Einstein, *Ideas and Opinions* (New York: Dell, 1976), p. 271.

[21]See Melic Capek, "Do the New Concepts of Space and Time Require a New Metaphysics," in *The World View of Contemporary Physics*, ed. Richard E. Kitchener (Albany, N.Y.: S.U.N.Y. Press, 1988), pp. 90–104.

[22]Henry P. Stapp, "S Matrix Interpretation of Quantum Theory," *Physical Review*, 1971, *3*, p. 1303ff.

[23]Henry J. Folse, "Complementarity and Space-Time Descriptions," in *Bell's Theorem, Quantum Theory and Conceptions of the Universe*, p. 258.

[24]Bertrand Russell to Gottlob Frege, 16 June 1902, in *From Russell to Gödel*, trans. and ed. Jean van Heijenoort (Cambridge, Mass.: Harvard University Press, 1967), p. 125.

[25]Gottlob Frege to Bertrand Russell, 22 June, 1902, Ibid, p. 127.

[26]Kurt Gödel, "On Formally Undecidable Propositions of *Principia Mathematica* and Similar Systems," in *From Frege to Gödel*, trans. and ed. Jean van Heijenoort (Cambridge, Mass.: Harvard University Press, 1967).

[27]See Ernst Nagel and James R. Newman, "Gödel's Proof," in Newman, ed., *The World View of Mathematics* (New York: Simon and Schuster, 1956), pp. 1668–1669 and Ernst Nagel and James R. Newman, *Gödel's Proof* (New York: New York University Press, 1958).

Chapter 5

[1]See Niels Bohr, *Atomic Physics and Human Knowledge*.

[2]Niels Bohr, "Biology and Atomic Physics," in Ibid., pp. 20–1.

[3]Niels Bohr, "Light and Life," in *Interrelations: The Biological and Physical Sciences*, ed. Robert Blackburn (Chicago: Scott Foresman, 1966), p. 112.

[4]Charles Darwin, "The Linnean Society Papers," in *Darwin: A Norton Critical Edition*, ed. Philip Appleman (New York: Norton, 1970), p. 83.

[5]Charles Darwin, *On the Origin of Species* (New York: Mentor, 1958), p. 75.

[6]Darwin, Ibid., p. 120.

[7]Ibid., p. 29.

[8]Lynn Margulis and Dorian Sagan, *Microcosmos: Four Billion Years from Our Microbial Ancestors* (New York: Simon and Schuster, 1986), p. 16.

[9]Ibid., p. 18.

[10]Ibid.

[11]Ibid., p. 19.

[12]Paul Weiss, "The Living System," in *Beyond Reductionism: New Perspectives in the Life Sciences*, ed. A. Koestler and J.R. Smythies (Boston: Beacon, 1964), p. 200.

[13]J. Shaxel, *Gruduz der Theorienbuldung in der Biologie* (Jena: Fisher, 1922), p. 308.

[14]Ludwig von Bertalanffy, *Modern Theories of Development: An Introduction to Theoretical Biology*, trans. J.H. Woodger (New York: Harper, 1960), p. 31.

[15]Ernst Mayr, *The Growth of Biological Thought: Diversity, Evolution and Inheritance* (Cambridge, Mass.: Harvard University Press, 1982), p. 63.

[16]P.B. Medawar and J.S. Medawar, *The Life Sciences: Current Ideas in Biology* (New York: Harper and Row, 1977), p. 165.

[17]Lynn Margulis and Dorian Sagan, *Microcosmos*, p. 265.

[18]Charles Darwin, *On the Origin of Species*, p. 83.

[19]Ibid., p. 77.

[20]Ibid., p. 75.

[21]Ibid., p. 76.

[22]Ibid., pp. 78–9.

[23]Richard M. Laws, "Experiences in the Study of Large Animals," in *Dynamics of Large Mammal Populations*, ed. Charles Fowler and Time Smith (New York: Wiley, 1981), p. 27.

[24]Charles Fowler, "Comparative Population Dynamics in Large Animals," in *Dynamics of Large Mammal Populations*, pp. 444–5.

[25]See David Kirk, ed., *Biology Today* (New York: Random House, 1975), p. 673.

[26]Charles Elton, *Animal Ecology* (London: Methuen, 1968), p. 119.

[27]David Lack, *The Natural Regulation of Animal Numbers* (Oxford: Oxford University Press, 1954), pp. 29–30, 46.

[28]V.C. Wynne-Edwards, "Self-Regulating Systems in Populations and Animals," *Science*, March 1965, *147*, p. 1543.

[29]James L. Gould, *Ethology: Mechanisms and Evolution of Behavior* (New York: Norton, 1982), p. 467.

[30]Paul Colinvaux, *Why Big Fierce Animals Are Rare: An Ecologist's Perspective* (Princeton, N.J.: Princeton University Press, 1978), p. 145.

[31]Ibid., p. 146.

[32]Peter Farb, *The Forest* (New York: Time-Life, 1969), p. 116.

[33]Peter Klopfer, *Habitats and Territories* (New York: Basic Books, 1969), p. 9.

[34]Eugene P. Odum, *Fundamentals of Ecology* (Philadelphia: Saunders, 1971), p. 216.

[35]Lynn Margulis, *Symbiosis in Cell Evolution* (San Francisco: Freeman, 1981), p. 163.

Chapter 6

[1]David N. Schramm, "The Early Universe and High-Energy Physics," *Physics Today*, April 1983, p. 27.

[2]G.F. Smoot in *Examining the Big Bang and Diffuse Background Radiations*, ed. M. Kafatos and Y. Kondo (Dordrecht, Holland: Kluwer Academic Publishers, 1986), p.31.

[3]See Report on Physics News, *Physics Today*, January 1988.

[4]R. Brent Tully, "More about Clustering on a Scale of 0.1c," *Astrophysical Journal*, 1987, *323*, pp. 1–18.

[5]Joseph Silk, talk presented at the Smithsonian *Is Cosmology Solved?* October 4, 1998.

[6]Alan H. Guth and Paul J. Steinhardt, "The Inflationary Universe," *Scientific American*, May 1984, p. 116.

[7]Ibid.

[8]Joseph Silk, 1998.

Chapter 7

[1]A.S. Eddington, *Monthly Notices of the Royal Astronomical Society*, 1931, *91*, p. 412.

[2]P.A.M. Dirac, *Nature*, 1937, *139*, p. 323 and Dirac, *Proceedings of the Royal Society*, 1938, *A165*, p. 199.

[3]F.J. Dyson, in *Aspects of Quantum Theory*, A. Salam and E.P. Wigner, eds. (Cambridge, England: Cambridge University Press, 1972).

[4]R.R. Harrison, *Cosmology: The Science of the Universe*, (Cambridge, England: Cambridge University Press, 1981), p. 329, and M. Kafatos, (1986), in *Astrophysics of Brown Dwarfs*, M. Kafatos, R.S. Harrington, and S.P. Maran, eds. (Cambridge, England: Cambridge University Press, 1986), p. 198.

[5]J.A. Wheeler, in *Some Strangeness in the Proportion*, ed. H. Woolf (Reading, Mass.: Addison-Wesley, 1981).

[6]M. Kafatos, in *Bell's Theorem, Quantum Theory and Conceptions of the Universe* (Dordrecht, Holland: Kluwer Academic Publishers, 1989), p. 195.

[7]J.A. Wheeler, in *Some Strangeness in the Proportion*, ed. H. Woolf (Reading, Mass.: Addison-Wesley, 1981).

[8]M. Kafatos and R. Nadeau, *The Conscious Universe: Part and Whole in Modern Physical Theory* (New York: Springer-Verlag, 1990).

[9]M. Kafatos, "The Universal Diagrams and the Life of the Universe," in *The Search for Extraterrestrial Life*, ed. Michael D. Pagagianis (Dordrecht, Holland: D. Reidel Co., 1998), pp. 245–249.

Chapter 8

[1]Henry P. Stapp, "Quantum Theory and the Physicist's Conception of Nature: Philosophical Implications of Bell's Theorem," in *The World View of Contemporary Physics*, ed. Richard E. Kitchener (Albany, N.Y.: S.U.N.Y. Press, 1988), p. 40.

[2]Ibid., p. 40.

[3]See Henry P. Stapp, "Quantum Ontologies," in *Bell's Theorem, Quantum Theory and Conceptions of the Universe* (Dordrecht, Holland: Kluwer Academic Publishers, 1989), pp. 269–278.

[4]See David Bohm, *Wholeness and the Implicate Order* (London: Routeledge and Kegan Paul, 1980).

[5]Ibid.

[6]Henry P. Stapp, "Quantum Ontologies," p. 273.

[7]Ibid.

[8]Ibid., p. 275.

[9]Henry P. Stapp, "Why Classical Mechanics Cannot Naturally Accommodate Consciousness But Quantum Mechanics Can," *Noetic Journal*, 1997, pp. 85–86.

[10]Roger Penrose, *Shadows of the Mind* (New York: Oxford University Press, 1994), p. 358.

[11]Ibid., p. 376.

[12]Roger Penrose, "Precis of *The Emperor's New Mind*: Concerning Computers, Minds, and the Laws of Physics," *Behavioral Sciences*, 1990, *13*, p. 643.

[13]David Bohm, "Interview," *Omni*, January, 1987, *9*, pp. 69–74.

Chapter 9

[1]Copernicus, Re Revolutionibus, quoted in Gerald Holton, *Thematic Origins of Modern Thought* (Cambridge, Mass.: Harvard University Press, 1974), p. 82.

[2]Kepler to Hewitt von Hohenberg, quoted in Ibid, p. 76.

[3]Galileo Galilei, quoted in Ibid., p. 307.

[4]Alexander Koyre, *Metaphysics and Measurement* (Cambridge, Mass: Harvard University Press, 1968), pp. 42–3.

[5]Heinrich Hertz, quoted in Heinz Pagels, *The Cosmic Code* (New York: Basic Books, 1983), p. 301.

[6]Ivor Leclerc, "The Relation Between Science and Metaphysics," *The World View of Contemporary Physics*, ed. Richard E. Kitchener (Albany, N.Y: S.U.N.Y. Press, 1988), p. 30.

[7]Ibid., p. 27.

[8]Ibid., p. 28.

[9]Ibid.

[10]Ibid.

[11]Ibid., p. 29.

[12]Ibid., p. 31.

[13]Ibid., pp. 25–37.

[14]Albert Einstein, *The World As I See It* (London: John Lane, 1935), p. 134.

[15]Ibid., p. 136.

[16]Ivor Leclerc, "The Relation between Natural Science and Metaphysics," p. 31.

[17]Albert Einstein, "Autobiographical Notes," in *Albert Einstein: Philosopher-Scientist*, ed. P. A. Schlipp (New York: Harper and Row, 1959), p. 210.

[18]Albert Einstein, "On the Method of Theoretical Physics," in *Ideas and Opinions* (New York: Dell, 1973), pp. 246–7.

[19]Ilse Rosenthal-Schneider, "Reminiscences of Conversations with Einstein," July 23, 1959, quoted in Ibid., p. 236.

[20]Gerald Holton, "Do Scientists Need Philosophy?," The Times Literary Supplement, November 2, 1984, pp. 1231–4.

[21]Henry P. Stapp, "Quantum Theory and the Physicist's Conception of Nature: Philosophical Implications of Bell's Theorem," in *The World View of Contemporary Physics*, p. 38.

[22]Ibid.

[23]Melic Capek, "New Concepts of Space and Time," in Ibid., p. 99.

[24]Henry P. Stapp, "Quantum Theory and the Physicist's Conception of Nature: Philosophical Implications of Bell's Theorem," in Ibid, p. 54.

[25]Werner Heisenberg, *Physics and Philosophy* (London: Faber, 1959), p. 96.

[26]Errol E. Harris, "Contemporary Physics and Dialectical Holism," in *The World View of Contemporary Physics*, p. 161.

[27]Ibid., pp. 162–3.

[28]Ibid.

[29]Ibid., p. 162.

[30]Ibid., p. 163.

[31]Ibid., p. 164.

[32]Ibid.

[33]Ibid., p. 165.

[34]Ibid., p. 171.

[35]Ibid.

[36]Wolfgang Pauli, in *Quantum Questions*, ed. Ken Wilbur (Boulder, Colo.: New Science Library, 1984), p. 163.

[37]Erwin Schrödinger, in *Quantum Questions*, p. 97.

[38]Erwin Schrödinger, in *Quantum Questions*, p. 81.

Appendix

[1]Menas Kafatos, "Horizons of Knowledge in Cosmology," in *Bell's Theorem, Quantum Theory and Conceptions of the Universe*, pp. 195–210.

[2]Ibid.

[3]See David N. Schramm, "The Early Universe and High-Energy Physics."

[4]George Miley, *Physics World*, 1989, *2*, p. 35.

[5]K.C. Chambers et al., *Nature*, 1987, *329*, p. 604.

[6]K.C. Chambers et al., *Astrophysical Journal (Letters)*, 1988, *329*, p. L75.

[7]Menas Kafatos, "Horizons of Knowledge in Cosmology," pp. 195–210.

[8]See, for example, the discussion in Edward R. Harrison, "The Cosmic Numbers," *Physics Today*, December 1972, p. 30.

[9]Anthony L. Peratt, *Science*, January/February 1990, p. 24.

[10]Halton Arp, *Astrophysical Journal (Letters)*, 1984, *277*, p. L27.

[11]Emil Wolf, see, for example, *Science News*, 1989, *136*, p. 326.

[12]S. Roy, M. Kafatos, and R. Dutta, *Physical Review A*, 1999 (in press).

[13]Menas Kafatos, "Horizons of Knowledge in Cosmology," p. 209.

Index